The Ocean World of Jacques Cousteau

A Sea of Legends: Inspiration from the Sea

The Ocean World of Jacques Cousteau

Volume 13

A Sea of Legends: Inspiration from the Sea

THE DANBURY PRESS

***The sea gives shape** to all it touches. So, too, the maker of myth and the teller of tales, the painter and the poet, give form to their art. The patterns of the sea have a significance that is physical or biological. The shape the artist gives his work has a significance that is aesthetic. Its concern is with the survival of the soul.*

The Danbury Press
A Division of Grolier Enterprises Inc.

Publisher: Robert B. Clarke

Production Supervision: William Frampton

Published by Harry N. Abrams, Inc.

Published exclusively in Canada by
Prentice-Hall of Canada, Ltd.

Revised edition—1975

Project Director: Steven Schepp

Managing Editor: Richard C. Murphy

Assistant Managing Editor: Christine Names
Senior Editor: Ralph Slayton
Editorial Assistant: Joanne Cozzi

Art Director and Designer: Gail Ash

Assistant to the Art Director: Martina Franz
Illustrations Editor: Howard Koslow

Creative Consultant: Milton Charles

Printed in the United States of America

234567899876

LIBRARY OF CONGRESS CATALOGING
IN PUBLICATION DATA

Cousteau, Jacques Yves.
A sea of legends.

(His The ocean world of Jacques Cousteau; v. 13)
1. Ocean. 2. Voyages and travels.
I. Title.
GC21.C684 551.4'6 74-23069
ISBN 0-8109-0587-6

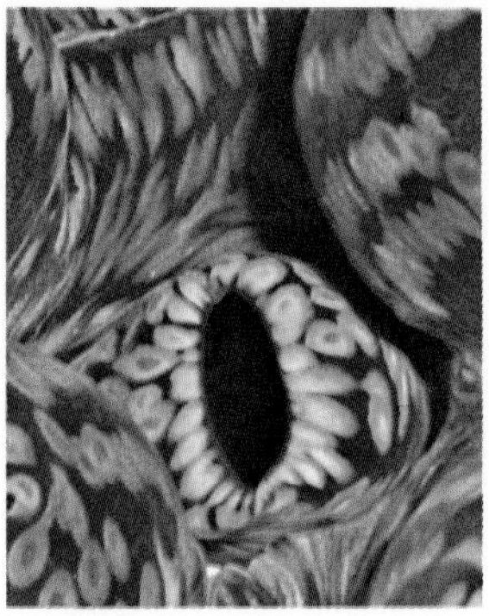

Contents

However alien the sea may have been to man in the past, she has influenced, directly or indirectly, all human activities. Today man is becoming ever more familiar with the great marine world and has penetrated to the deepest abyss. Everyday the sea inspires more profoundly all arts, sciences, and even our way of thinking. We should get more familiar with the WET MUSES.

Once the sea was only a dark, fearsome expanse beyond man's ken. When he asked questions about it, as he did about everything in nature, he found many of the answers in his imagination. The myths that he invented not only provided explanations but also served to make the terrifying ocean somewhat less threatening. He asked how THE WATERS UNDER THE HEAVEN (Chapter I) had been made, and each society found a different answer. He asked how the fish and all the other animals of the sea had been made and why the tides ebbed and flowed. He asked if there had not been whole cities or even continents that had been swallowed by the sea.

Man peopled the sea with gods and goddesses. They could be supplicated and the sea made less terrible. The ocean itself he made a god and all its life became sacred. Then as his religions developed, these gods and goddesses were inextricably bound up in superstition and reality.

Quite early man discovered that it was easier to travel along the rivers than to go by land; waterways allowed a broader exchange of goods and ideas, and civilizations grew. All these rivers led to the sea, and one day man accepted THE CHALLENGE (Chapter II) and ventured beyond the estuary onto the dangerous ocean. He came back with new myths.

Overland travelers returned from distant parts of the world with fabulous stories of great wealth and with jewels and spices to prove it. The sea could provide an easy route to these treasures if only a way could be found. Man discovered, too, that the sea itself was a GOLCONDA (Chapter III) no less than the fabled cities of diamonds. The greed to possess its riches showed itself early and never abated. Inevitably the rush to acquire this wealth led to conflict, and there were very few laws to govern the use of the sea. It soon became apparent that whichever nation controlled the sea might control the land, and one nation after another came to hold the hegemony.

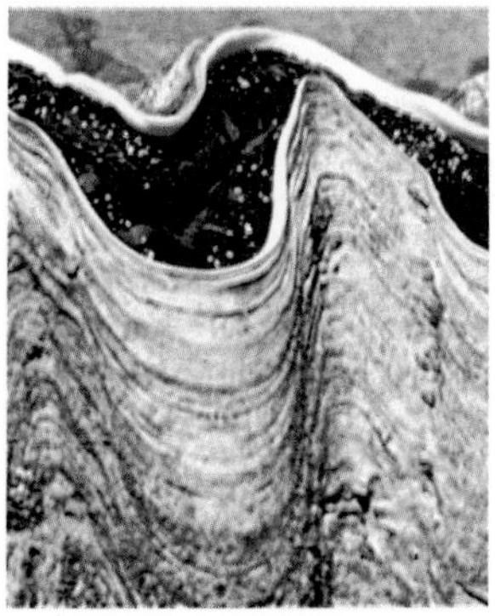

More and more men followed the challengers, many of them prompted by boredom and restlessness as well as by the lure of wealth, some of them simply curious to know what life was like ELSEWHERE (Chapter IV). In the West, the Norsemen, who had been compelled by need to dare the sea, became the first modern explorers in ships. They even discovered America, but they did not bother to tell others about it, and the rest of the world remained ignorant of their accomplishment for centuries. Eventually others joined them, and a period of rapid development of the naval arts opened the way to what would come to be called the Great Age of Discovery—the age of Diaz and da Gama, of Columbus, Balboa, Magellan, and many other brave explorers. Sailors returned from these voyages with new tales to tell and with elaborations on old ones. Many of their stories were of frightful monsters. 64

The intellectual awakening that took place in the West at this time, which we call the Renaissance, prompted a search for new knowledge in what were to become the sciences of the sea. The greatest mind of the time—Leonardo da Vinci—turned his attention for a time to trying to find a way to permit man to explore the sea beneath the surface. As man achieved a better understanding of the oceans, he regarded it at times with something approaching delight. He conceived a charming inhabitant of the watery realm—the mermaid—who was to make many a long voyage a little less arduous. THE BIRTH OF VENUS (Chapter V) took place in the sea. 86

But that is not to say that the sea had been tamed. Far from it. In reality she was as dangerous as ever. No one could love the sea without hating her at the same time. Hadn't the sea taught the sailor to curse? The sea was THE IMPOSSIBLE MISTRESS (Chapter VI). 110

The sea that had once belonged to no one eventually became the concern of all. The sea that no one had given any thought to caring for was discovered to be essential to man's survival. The struggle to possess it that had begun so long ago grew progressively more fierce. The international law of the sea such as there was had not sufficed. It would be necessary to pass laws of international cooperation if the necessary RECONCILIATION (Chapter VII) were to be achieved and if man were to live wisely and well with the sea. 132

The carnal bonds that have kept salt from the ocean coursing in our bloodstream today lure us to share the SPIRIT OF THE SEA. 142

Introduction: Wet Muses

In the clear tropical waters of the Red Sea, along the exuberant coral reef of Port Sudan, a great iron ship ran aground and sank many years ago. It has since been coated with a medley of graceful polyps, sponges, and gorgonians; it was invaded by scores of gaudy reef fish, feather-duster worms, starfish, and sea snails that surpass in brightness and gorgeous colors the butterflies and flowers of our gardens. On board *Calypso,* André Laban dons his diving equipment and slowly walks down the ladder into the sea, his arms burdened with unusual impediments: a tripod, a metal stool, a large rectangular canvas, and a bag full of brushes and paint tubes. He is heavily weighted and sinks rapidly to an appropriate site he had chosen during previous dives. Soon he is comfortably installed, facing the wreck's crumbling and festooned bridge, and he begins to paint the scene. He then looks like one of the Sunday artists exercising their talents in front of Notre Dame in Paris. The difference is that this underwater painter releases streams of silver bubbles. Back on deck, Laban carefully flushes his "masterpiece" with fresh water from a hose. The first known undersea artist was Pritchard, who painted in a heavy helmet suit at the end of the nineteenth century.

Painting was the first—and the most obvious—of the arts to be influenced by the sea. Photography followed, as soon as the peculiar technical problems raised by the physical and optical properties of seawater had been solved. Today it is not enough to make a well-exposed, color-corrected, well-focused picture: the art of photography also implies tasteful composition and the choice of the most effective angles.

Off the coast of the Italian Riviera, a large bronze Christ has been installed in the depths, and a statue of the Virgin watches over the undersea world in Spain. Miró made a ceramic sculpture of a strange creature meant to be the Spirit of the Deep, which was set up in a cave, 120 feet down in the Mediterranean. Aside from these spectacular pieces, the art of sculpture is often inspired by such streamlined shapes as the fluid, graceful lines of the dolphin.

Architecture is also faced with new problems. Engineers had early solved the problems of designing structures to withstand water pressure. For functional reasons hulls or habitats intended for underwater use were either spherical, cylindrical, or ellipsoidal, but they had to be tailored to the needs of men, and until recently, the absence of specialized diving architects has badly handicapped their efficiency. Conshelf II and Conshelf III were the first to be influenced by architects, designers, and decorators, as well as by engineers. Undersea settlements of the future will certainly be drafted by teams of architects and technicians such as those that are responsible for the most beautiful and efficient buildings on land.

Poetry and music have hitherto been very slightly influenced by the sea, and most generally only by her surface aspects: storm, doldrums, heroism of the navigators, destination of the fishermen. But poets and composers are today discovering new sources of motivation in the realities of the undersea world. A full symphony inspired by the melodious songs of the humpback whales recently marked the beginning of a new field of music.

Writers and playwrights have long been fascinated by the mysteries of the ocean and by information available in the legends of the sea. But today facts are just as extraordinary as tales, and quality books of science fiction or science anticipation are written in increasing numbers.

The popular films of the aquatic ballets of Esther Williams were an early attempt at including the sea in some aspects of the art of dancing. But there remains a wide fascinating field to be explored. All the theory and the mathematics of choreography are based on the inescapable forces of gravity: humans are prisoners of their weight and thus sentenced to move around and struggle in a two-dimensional world. Dance expresses the efforts of the body and the dreams of the mind to escape gravity. Weightlessness, experienced by divers in their three-dimensional realm, could be a totally new departure for a different choreography.

Inspirations from the sea are not limited to the arts. In the field of exact sciences, oceanography is rapidly developing, not so much as a new science, but as the field for the study of a new environment, and this endeavor includes, of course, all the exciting scientific specialties.

A more subtle but overwhelming influence is rapidly spreading and slowly penetrating our daily life. Studies of the behavior of marine animals show that even in such a different environment as the sea, the motivations of all creatures are basically the same. The concept of the unity of life is every day more obvious. The oceans, in the same manner as space, are contributing to link the nations of the world; oceanographers of all countries work hand in hand in an extraordinary atmosphere of understanding and collaboration. The obsolete laws governing the high seas and the exploitation of all provinces of the ocean are about to be revised, and new rules will establish that the seas are the common property of mankind. The damages caused to the world by the mismanagement and pollution of the sea call for action on a global scale and in a way will help a stronger solidarity of our species. In fact, our entire philosophy is about to be changed by the sea.

Jacques-Yves Cousteau

Chapter I. The Waters Under the Heaven

In the Revelation of Saint John the Divine, the last book of the Bible, John imagines a utopia in which there is earth and sky but no sea: "Then I saw a new heaven and a new earth, for the first heaven and the first earth vanished, and there was no longer any sea."

"In the beginning God created the heaven and the earth. And the earth was without form, and void; and darkness was upon the face of the deep."

John's fear of the sea was what every man felt. The sea was great, and it was a stranger. It was said that it led to the edge of the world, perhaps to the land of the dead. This fear inspired the same response as did all natural phenomena—man created a mythology to answer the unanswerable, to explain the inexplicable, and to assuage, a little, his terror.

There was the very creation of the seas to be explained. In the Bible, God created the sea when he made the heavens and the earth. In Teutonic mythology the ocean was created from the blood of a giant who was the first living thing to emerge from the emptiness that lay between the land of fire and the land of ice and mist. In Egyptian mythology a great ocean existed before the creation of the earth and the heavens.

It was usually in terms of gods and goddesses and heroes that early man faced his fears in his myths. The underlying theme running through the early myths of the sea was the idea of its being a source of divine power for gods and kings and later for saints.

The early myths played an important role in each society's religious life, and they were usually considered to be sacred and true. The divine beings with which most of the myths are concerned have not only supernatural powers but human characteristics as well. They have emotions like love and jealousy, and they experience birth and death. By peopling the sea with gods who resembled himself, man was in effect endowing himself with a knowledge of it and a power over it that he did not possess.

The Greek and Roman gods of the sea were to give ground to Christianity, but their echoes would remain. Some of the old rites were incorporated into the new religion.

Fish early appeared in myth and religion. The Carthaginians worshipped the fish as a symbol of strength. The Egyptians regarded eels as sacred. The Chaldeans worshipped the fish-god Cannes, who was believed to have bestowed the gift of wisdom on the first humans. In Peru the Yuncas adored the ocean in the form of a fish out of gratitude for the food it yielded.

The sea was indeed like a mother, and man himself was in his childhood. The day would come when he would seek to free himself and to set out to discover the world on his own, and the sea would become a young woman for him. He would take great delight in her but he would find out, too, that she could be fickle, even treacherous. One day he would come to accept her for what she really is, and he would begin to care for her as she cared for him.

Oceanus, *in Pelasgian myth, was created by the goddess of all things, Eurynome. To later people he was the earth-girdling stream in which all gods and living creatures originated. Oceanus and his Titan queen Tethys were parents of 6000 children—all the waves of the ocean—and ruled over earth's waters until they were supplanted by the Olympians Poseidon and Amphitrite.*

The Flood and the Covenant

Many stories of great floods occur in the mythology of almost every society, and they are remarkably similar. What inspired them? Perhaps it was a real deluge, but it seems more likely that because the sea was the most dreaded of natural phenomena its wrath became the instrument of the worst disaster man could imagine.

The most familiar flood story is probably that of Noah and the Ark. The essence of that story, found in the Old Testament, is that the people had multiplied and become more and more wicked. God was sorry that he had created man, and he decided to destroy him. There was one man, however, who was good. His name was Noah, and he had a wife and three sons. God commanded him to build a ship and told him that soon he would cause the rain to fall until the whole earth was flooded and every living thing was destroyed. God further told Noah that he was to take into the ship himself, his wife, his three sons and their wives, and two of every other living creature—a male and a female of every kind of bird and animal.

Noah's neighbors jeered at him when they saw him building a boat in the middle of a dry field, and they thought he had gone mad. But Noah did not care what the neighbors thought, for he was doing the will of God.

When the ark was finished, God told Noah to start loading it. Two by two the creatures came—a pair of every sort of bird and beast that lived—and they went on board the ship. Then Noah and his wife and three sons and their wives went into the ship. Seven days later it began to rain, and it continued until the ark was lifted up onto the water. And still it rained, and the waters rose so high that even the tallest mountains were covered. Everything that moved was drowned except for the company of the sturdy ship.

After 40 days and nights the rain stopped and slowly the waters began to recede. At last the ark came to rest on the top of a mountain said to be Mount Ararat in eastern Turkey. Presently the tops of the mountains were visible. Noah opened a tiny window. All around him spread the endless water. He sent forth a raven, but the bird could only fly back and forth, to and fro. Then Noah sent a dove forth, but the dove, like the raven, found no dry ground on which to light, and the bird returned to the ark.

Noah waited another seven days, and then he again released the dove. In the evening the dove came back. In its mouth was a freshly plucked olive branch. So Noah knew that the waters had gone down.

Now God commanded Noah and all the company of the ship to go forth and to be fruitful and multiply. And he made a promise to Noah that he would never again send a flood to destroy mankind, and he put a rainbow in the sky as a symbol of his promise. Henceforth, man could live at peace with the sea.

Noah frees the dove *in a mosaic from Saint Mark's Church in Venice dating from the twelfth century. Noah opened the window of the ark:*

And he sent forth a raven, which went forth to and fro, until the waters were dried up from off the earth.

Also he sent forth a dove from him, to see if the waters were abated from off the face of the ground.

But the dove found no rest for the sole of her foot, and she returned unto him into the ark, for the waters were on the face of the whole earth: then he put forth his hand, and took her, and pulled her in unto him into the ark.

GENESIS

FORTI
QVEM · CAP · VDA · FRETI · COCLVDVNT · VISCERA · CETI ·
IONAS · I VENTRE · IET

"Me?" Said the Whale

When God needed an errand done, he called upon the whale. He stood at the shores of the sea and with the voice of a storm summoned the whale that rose gleaming from the sea and swam obediently before its God. The Lord spoke, and the whale listened.

"I see that my prophet Jonah will not go to Nineveh to preach as I shall command. He will attempt to flee from me, and he will choose the sea as the best means of escape. It will not help him. I will raise a great storm on the waters, and the sailors will throw him overboard as a sacrifice to the gods of the storm. This is where you come in, my old friend. As he sinks through the water, you will swallow him."

"An unpleasant task," the whale mused, and he heaved a sigh that caused all the little fish in the sea to tremble.

"Perhaps," said God, "but just think. It will give you a place in history." Then he blessed his servant and the whale sank sadly back into the depths of the sea.

Jonah did indeed refuse to go to Nineveh as God commanded. He might as well preach to the fish of the sea, he thought to himself. Instead of doing what God told him to do, he boarded a ship bound for Tarshish. The sea was calm when they set out, and Jonah curled up in a corner of the deck and soon he was sound asleep. In the afternoon an ominous haze began to cloak the sky, and the sea became oily. In the east it grew darker and darker. All at once the sky groaned and the waters began to move. And then it seemed as if the sky fell down from heaven and the sea rose up to meet it. Panic-stricken, the sailors rushed to throw their cargo overboard. As they went about their work they prayed. But Jonah refused to pray.

The sailors knew that the storm was not an ordinary one, but a storm of the gods. Someone had made a god very angry indeed. It could only be the stranger who seemed to fear neither god nor storm, and although they were good men who did not want to take a life, they threw Jonah overboard because they believed it to be God's will. At once the sea grew calm, the wind died down, and the sun sank tranquilly in the west.

Jonah sank through the water without any effort to save himself. It was the end and he had no desire to live. But as his breath failed, he began to remember the blue and shining sky, the sweet odors of the desert, and the happy dreams of his childhood. And he began to struggle for his life.

Suddenly he found himself in a deep, dark cavern. The whale had done as he had been told and had swallowed Jonah. The whale seemed to speak to Jonah, but it was God himself. "I am your God, Jonah, and where you go, there you will find me." Jonah thanked God with all his heart for saving him, and he promised never again to refuse to do his bidding. God was satisfied, and the whale, lashing the waters with his tail, sped to the shore and spit Jonah up upon the sand.

Jonah *is thrown overboard and the whale waits to swallow him in this altarpiece from Verdun (left).*

The Shaker of the Earth

The powerful god of the sea, Poseidon, is believed to have come to the Greek peninsula as the chief diety of northern Aryan tribes that settled near the Mediterranean around 2000 B.C. He had created horses for them, and because the stampeding animals caused the earth to tremble, he was also identified with earthquakes. As a symbol of his dominion over the seas he usually carried a trident, which he used to smite the earth to cause it to quake or to bring forth fountains and horses.

When the Greeks named Zeus as the first god of Olympus and ruler of the sky, his elder brother, Poseidon, was assigned command of the sea and all that belonged to it. He lived in a palace deep in the lagoon with his wife Amphitrite, eldest of the Nereids—daughters of an earlier sea god, Nereus. Homer's description of Poseidon revealed him standing upon his chariot, clothed in gold with a golden whip. His golden-maned horses bounded across the waves while all the sea monsters gamboled at the coming of their king.

Poseidon could be a formidable enemy. His anger at the Trojan king Laomedon, who failed to pay a promised reward to him and Apollo, caused him to send a terrible monster to ravage the city. He favored Troy's enemies, the Greeks, but later he turned against the arrogant victors and made their return home a disaster. Storms destroyed many of their ships and blew others far off their course. But the great god of the sea could be kind as well. He was the protector of seafarers and fishermen. His sacred dolphin was a symbol of a calm and peaceful sea. He was worshipped by all people who dwelt by the water, and a festival in his honor was held in alternate years at the Isthmus of Corinth. Poseidon's statue could be found on many high places overlooking the sea.

From early times the Romans worshipped Neptune as the god of moisture and flowing water. Later they identified Neptune with Poseidon and when they did they recognized Neptune as god of the sea.

The god of the sea was called ***Poseidon*** *by the Greeks, Neptune by the Romans. As in the mosaic (left), he was often portrayed naked and forbidding, his hair and beard caked with salt and blown by the wind. He carried a trident, the symbol of his dominion over the waves.*

Poseidon's temple *(right) stands high above the sea at Cape Sounion, welcoming returning seamen home. Built on the ruins of an earlier monument destroyed by the Persians, this Doric temple dates from the time of Pericles, around 540* B.C.

Musician and Dolphin

Sunlight danced on the waters, soft breezes filled the sail of the ship bearing a triumphant Arion to his home in Corinth. He was pleased that the dire predictions of his friend, King Periander, had not come true. Arion had made the dangerous voyage to Sicily to enter a contest for the world's finest musician. That title and its rich reward were his. How they would celebrate on his return!

Suddenly the musician was surrounded by the crew. "Arion, you must die!" they shouted. "No", Arion pleaded, "take my gold, but spare my life." "You must die", the crew repeated. "We would not escape the wrath of Periander if we allowed you to live." "Then let me change my clothes and sing one last song that I may die as a bard." The request was granted. Arion dressed in a tunic of gold and purple, adorned his arms with jewels, and placed a wreath above his flowing scented hair. Here was a minstrel Apollo could approve. Then calling upon the Nereids to receive him, he jumped into the sea.

Arion's melody had attracted appreciative dolphins to the ship. One approached the musician as he fought to stay afloat and offered a ride on his back. He swam swiftly to a rocky shore where Arion dismounted. Bidding his benefactor good-bye, Arion expressed his regret that the dolphin could never be his companion on land, or he the dolphin's in the sea. Then Arion set off for his home, playing his lyre and singing.

*Shown in the picture left, the merman **Triton,** accompanied by a Nereid, was the son of Poseidon and Amphitrite. His torso was that of a man, but he had the tail of a fish. By blowing on his twisted snail shell, he was to quiet the waves.*

Fish symbol *(right) from a fifth-century Salonika church. In the Greek language an acronym from the words "Jesus Christ, Son of God, Savior" spells the word for fish.*

When the ship arrived in Corinth, its homicidal sailors told King Periander that they had left Arion in Tarentum, well and prosperous. At that very moment, to their amazement Arion appeared before them, dressed as he had been when they last saw him. The king announced a stiff punishment for the would-be killers: they were forever banished from his kingdom. On the rocky shore where Arion had landed, he and King Periander placed a brass monument in gratitude to the dolphin.

"The Battle of the Sea Gods," *an engraving by Andrea Mantegna (1431–1506). The Renaissance was a time of rediscovery of classical myths, largely forgotten during the Middle Ages.*

Children of the Sea

The ancient Greeks had a delightful imagination, and some of the beings with whom they peopled the sea could only have come from an affectionate regard for it.

The sea nymphs, as the early Greeks conceived them, were entirely human and feminine, from their shell-decked tresses to the soles of their feet. They were divided into the Nereids, who dwelt in the Inland Sea, and the Oceanids, who lived in the ocean.

The Nereids, 50 in number, lived in the Mediterranean and, in particular, the Aegean Sea. Probably the three best-known Nereids were Amphitrite, who married Poseidon and thus became Queen of the Sea, Thetis, reluctant bride of Peleus, and Galatea, beloved of the cyclops Polyphemus.

Nymphs of springs, ponds, *lakes, and rivers were called Naiades and were equivalent to the Nereids who presided over the sea. Nymphs were associated with all of nature's growing things.*

These nymphs had lovely musical voices. They danced around their father Nereus and sang in unison. Apart from dancing attendance upon Nereus, they waited upon the other powerful sea deities, and they owed absolute obedience to Poseidon. But they were minor deities in their own right; altars were erected in their honor along the coast, and offerings were made to them.

The Oceanids numbered no less than 3000. They were the daughters of Oceanus and Tethys and were similar in disposition to the Nereids. Like them, they were the recipients of offerings and prayers for safety amid the perils of the sea. They regarded sailors as being in their care, not as potential victims. One of them cared so much for a certain famous sailor that she would not let him go. Calypso, who reigned over the island of Ogygia, held Odysseus prisoner for seven out of the ten years it took him to find his way home to Ithaca. Calypso loved him and wanted him to stay and be her husband, and she promised to make him immortal, but Odysseus refused.

Greek legends tell us that not only were the Inland Sea and the ocean graced by Nereids and Oceanids, but that every river, spring, and fountain had its presiding maid. The maids were young and lovely virginal nymphs who were found in river, spring, and stream, as well as in the woods and meadows near the waters over which they presided. The Greeks held them in considerable veneration and poured libations of wine, oil, and honey to their honor.

The Lost Continent

When Solon, the Athenian statesman and poet, was on a trip to Egypt, he heard from priests there a tale that he used as the basis for a poem. Many years later Plato took up the tale and cast it in his own terms in his *Critias* and *Timaeus.* These dialogues tell one of the most enduring legends of all time —that of Atlantis, the lost continent.

Plato described an empire established "beyond the column of Hercules" (Gibraltar) and "larger than Egypt and Lybia," that was fabulously wealthy, based on native resources, imports from abroad, and an abundance of metals. There were found wild and domesticated animals, among them many elephants. The people of Atlantis had built temples, palaces, docks, a magnificent harbor, naval storehouses, and a system of canals and covered bridges. The capital was surrounded by a triple stone wall. In its citadel were a temple to Poseidon, several other temples, gardens, and the royal palace. In every way, it was a superior civilization.

But in a day and a night, according to Plato, floods and earthquakes struck Atlantis, and the island was swallowed up by the sea and vanished forever.

Ever since then people have wondered where Atlantis might have been. It has most popularly been thought to have existed somewhere in the Atlantic Ocean. Some have surmised that it was the New World. Many of those who would place the lost continent in the Atlantic believe that there might once have been a land bridge there. They were encouraged in the nineteenth century by the discovery of the Mid-Atlantic Ridge, a huge submarine range running roughly north-south along the ocean floor. It is now known that the Mid-Atlantic Ridge is not an isolated phenomenon but part of a belt of suboceanic ridges that extend well beyond the Atlantic Ocean. Another theory claims that

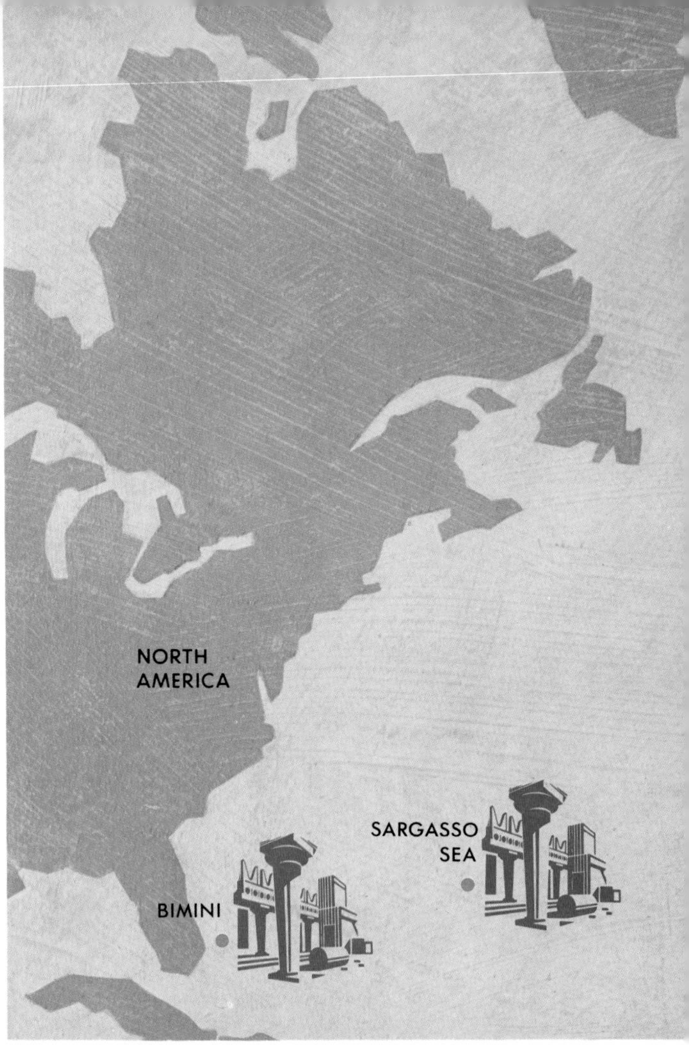

the Sargasso Sea in the North Atlantic lies above Atlantis and that the eels that go to this area from Europe and America to lay their eggs and die are in fact returning to an ancient homeland.

A number of interesting coincidences are presented in the legends of North and South America. The majority of Indian tribes have tales telling that they came from the east or that they obtained the arts of civilization from men who came from a continent to the east. The Aztecs preserved the name of their land of origin—Aztlan. In fact, the name Aztec is itself derived from Aztlan Quetzalcoatl, the god of the Aztecs and other Mexican peoples. He was said to be a bearded white man who came to the Valley of Mexico from the ocean. In their sacred book, the Quiche Maya refer to the paradisiac eastern country where they once lived until the god Hurakan (Hurricane) became angry and sent a great flood. When the Spaniards first

Countless theories have been advanced to prove the location of ***the lost continent of Atlantis.*** *Some of the sites more frequently mentioned are shown on this map.*

explored Venezuela, they found a settlement called Atlan, peopled by white Indians, who claimed that their ancestors had been survivors of a drowned empire.

Many other sites have been suggested for Atlantis, among them Palestine, Sweden, the region between Ireland and Brittany, the Helgoland region of the North Sea, the Arctic, Central Asia, Nigeria, Tunisia, and Spain. Some people think they have convincing evidence that it was the island of Thera in the Aegean Sea. Others insist that it was at Bimini, near Florida. An expedition to the Greek island, which lies directly north of Crete, has focused considerable attention on the theory that Thera itself, which apparently was destroyed in 1500 B.C. with the resultant submergence of a large land area, was the site of the actual disaster that gave Plato the idea of the destruction of a continent. It is known that a mysterious disaster struck the advanced Cretan civilization at about the same time, but evidence for the kind of civilization Plato described is extremely flimsy.

Near Bimini several buildings seem to be rising to the surface. They may not be buildings at all, however, but merely geological structures with a geometric configuration. If, indeed, they are buildings, they might very well simply be structures of the Mayas or of other Indians with whom we are already familiar.

Probably we will never know what these structures are for certain. Perhaps the whole story was so embellished by Plato that what we have been seeking is no more than the invention of a philosopher and poet.

Winged cupid *rides a pair of racing dolphins in a mosaic floor of the late second century* B.C.*, part of the House of the Dolphins on Delos Island.*

A Boy's Best Friend

Today many a schoolboy would like to come out of school everyday to find a dolphin, his best friend, waiting to take him home on his back. Once, very long ago, this really happened. At least, that is what some of the authors of ancient times assert.

If we are to believe those writers, it is certainly true that a boy's best friend was a dolphin. According to Aristotle, the great Greek philosopher, "among the sea fishes many stories are told about the dolphin, indicative of his gentle and kindly nature, and of manifestations of passionate attachment to boys, in and about Tarentum, Caria, and other places." Somewhat later, the Roman historian Pliny related how "a dolphin that had been brought into the Lucrine Lake fell marvelously in love with a certain boy, a poor man's son." The boy and the dolphin became fast friends and "when the boy called to it at whatever time of day, although it was concealed in hiding, it used to fly to him out of the depth, eat out of his hand, and let him mount on its back, sheathing as it were the prickles of its fins, and used to carry him when mounted right across the bay to Pozzuoli to school, bringing him back in similar manner, for several years, until the boy died of disease, and then it used to keep coming sorrowfully and like a mourner to the customary place, and itself also expired, quite undoubtedly from longing."

In ancient legends dolphins were not only boys' best friends but guides for sailors lost at sea. Dolphins understand men because, according to legend, they were once men themselves. The gods, it was believed, had transformed them into dolphins. In the *Homeric Hymns* we find just such a legend. The child god Dionysus was captured by pirates of the Tyrrhenian Sea. Thinking that they would sell the boy, they put him on a ship to take him to the slave market. But the child was possessed with divine powers, and he began to transform the ship. He made red wine run over the decks and leaves sprout from the mast. He transformed himself into a lion. The pirates were terrified and leapt into the sea where they were immediately changed into dolphins. Not only was their form changed, but their character as well. Henceforth they began to aid mankind.

It is of little wonder then that there were moral admonitions against the killing of dolphins. Oppian, the Greek poet, declared that "the hunting of dolphins is immoral" and that the men who killed dolphins "would not spare their children or their fathers and would lightly slay their brothers born." Sailors today believe that dolphins will not play about a ship that has caused the death of one of their kind. The dolphins can tell, they say, and they stay away.

Playful companions. *Dolphins and cherubs representing the mischievous god Eros frolic on a tile floor in a pattern popular in Hellenistic Greece.*

The Sea That Roars

One of the most dangerous passages for the ships of old was through the Strait of Magellan at the southern tip of South America. An ancient myth known to the Yamana Indians of Tierra del Fuego helps us to understand the terror that those waters held.

The myth relates that in very ancient times women commanded men. The men were not at all happy with this arrangement because they were obliged to do all the housework, to keep the fires, and to clean the hides for tanning. Since the women knew that their power would one day be taken from them, they attributed that power to Tanuwa, the supreme goddess. They constructed a sacred dwelling where they met to carry out the most complicated initiation rites. Then, to further convince the men of their right to rule, they disguised themselves with masks and painted their bodies and presented themselves as incarnations of the goddess.

At this time the sun was not yet in the sky but lived with the men of the island. One day, when the sun was taking a walk in the woods, he discovered the women's secret, and he hurried to unveil their deceit to the male members of the tribe. The men were furious. They armed themselves with arrows, harpoons, and lances, and they marched to the sacred spot. In the struggle that ensued between the men and the women, the women were transformed into sea animals. The sun threw a pail of water on the temple, and it turned into a gigantic wave that carried off all the animals. From that day on, the sea roars when it gets angry.

The sea that roars. *Among the Yamana Indians of Tierra del Fuego, it is believed that the sea roars when it gets angry because women were once transformed into sea animals.*

The Ocean of the Cuña Indians

Among the many stories of the creation of the ocean to be found around the world is this one from the Cuña Indians of Panama.

Ibelelus, who was their demigod, saw a woman pass by, chanting these words: "Tree of salt . . . tree of salt . . ." Ibelelus followed her to see where she came from.

The next day he came back to his men and said: "I have discovered where the woman comes from. She lives in a tree of salt. At the top there is water and fish and flowers and other animals. At the foot of the tree there is a lot of gold." His nephews decided to go and attack the big tree, and they set to work the following day. But they did not even manage to cut the bark off. Ibelelus was surprised at this, and in the night he posted himself in the brush to find out who was keeping the tree from falling. He saw a jaguar, a snake, and a frog come to lick the trunk's wounds, which immediately healed. He resolved to get rid of these magic animals, and the following morning his nephews began again to cut the tree. The chips flew, and when they touched the ground, they were transformed into fish; but the tree remained standing because the clouds themselves supported it. Then Ibelelus commanded the squirrel to cut the clouds that held the foliage of the tree, and at last the tree fell. When it did, the water that was at the top spread everywhere and formed the sea.

Pegasus *(above), the winged horse of myth, sprang from the body of Medusa. When Pegasus stamped his hoof, the fountain of the Muses gushed forth.*

The ***seahorse*** *(right) became a minor Christian symbol in at least one part of the world. Crosses bearing its image are found in eastern Scotland.*

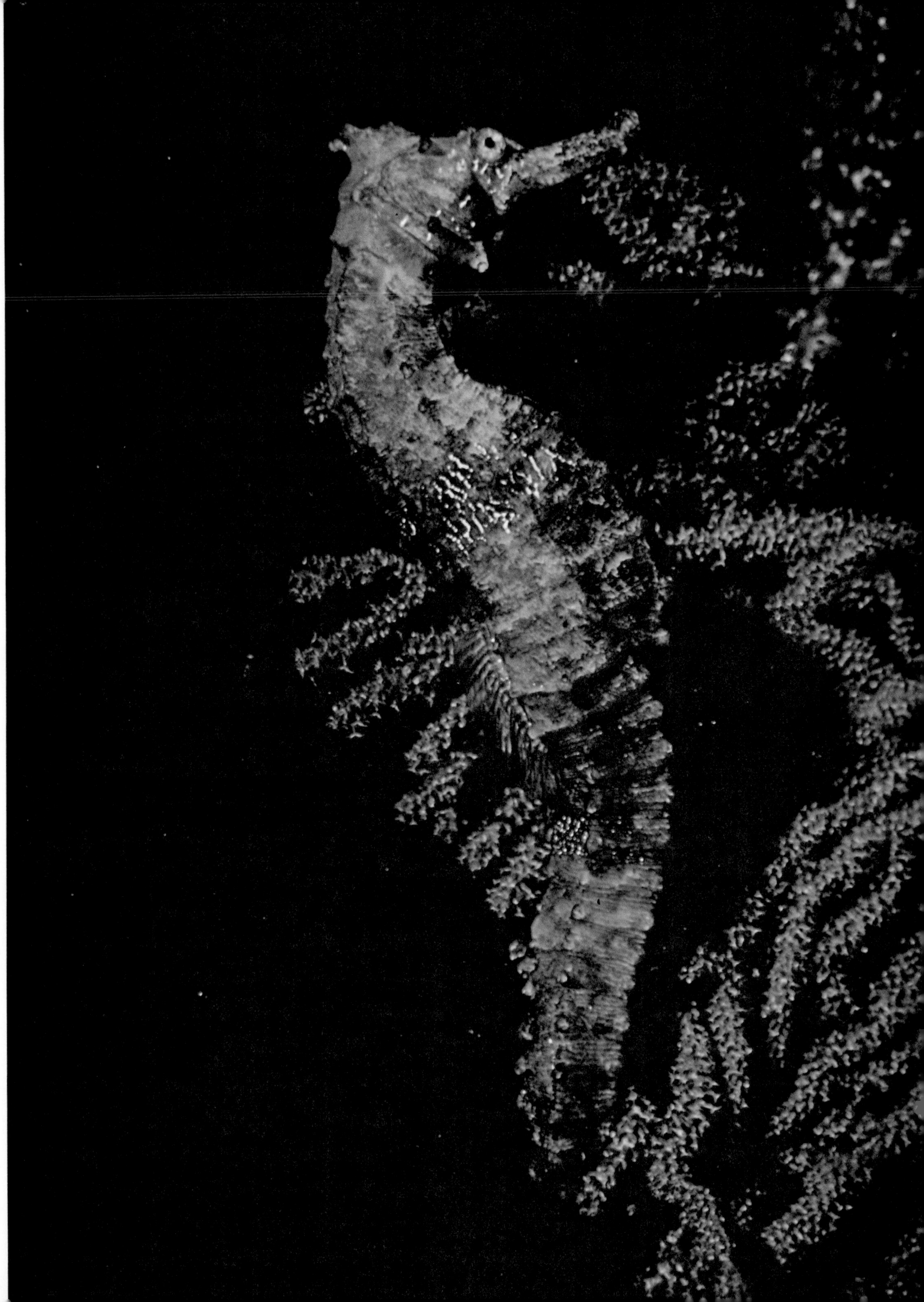

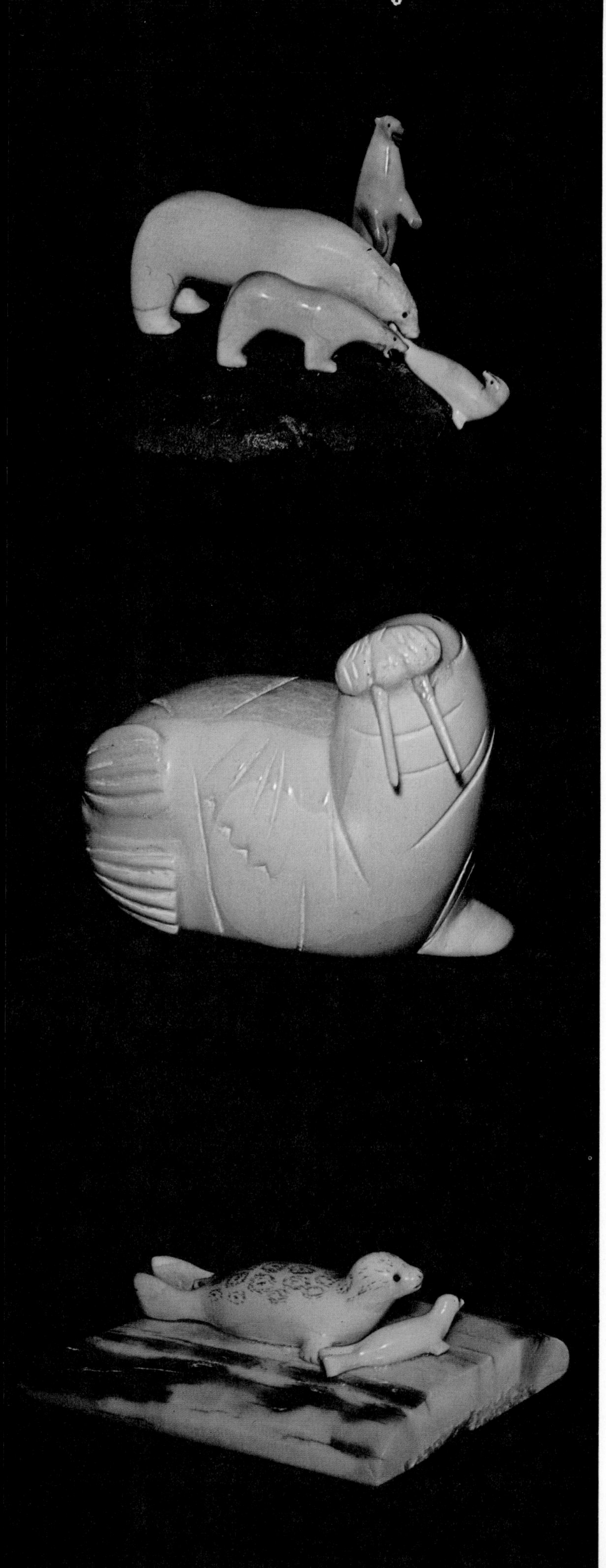

Sea Goddess of the Eskimos

Every part of the world has its legends of life in the sea. The Eskimos tell of a girl who became Sedna, the goddess of the sea. She was a girl who refused her suitors and married a bird. The girl's father, outraged at this, killed her husband and took her home in his boat. On the way, a storm arose, and the father threw the girl overboard. She clung to the gunwale until he chopped off her fingers, one by one. She then sank to the bottom of the sea, and there she still dwells, keeping guard over all that dwell there.

But her severed fingers were transformed into the fish and mammals of the sea, and these, her children, ate up the girl's father. The girl became the chief deity of the lower world, and each autumn the people of the Far North hold a great feast and festival in her honor. To some of the Eskimos she is known as Sedna, but she goes by a great number of other names in various parts of the arctic, and her story has many variants.

Nowadays her companions are a dwarf and an armless woman with whom she shares her husband, a sea scorpion.

Sedna feels no kindness towards humanity, but she does not act arbitrarily. She never moves about of her own free will but is rooted to her stone dwelling. Her sinister appearance would kill an ordinary man, and only a shaman, or priest, is able to stand the sight of her. Huge, voracious, and impotent, with a wild temper, she keeps watch with her one eye over the mammals of the sea.

When a hunter kills a mammal of the sea unnecessarily, Sedna feels resentment and great physical pain in the place where that animal originally sprang from her body.

Eskimo art. *Carvings are made of walrus tusk ivory (left) and soapstone (bottom right). Hunting scene (above right) is drawn on seal skin.*

Eskimo tale. *The Eskimos tell a tale of the origin of the narwhal. One day, so the legend goes, a boy and his mother went hunting white whales. The whales did not like being hunted, and one of them swam up behind her and pulled her into the sea. Terrified, the woman took her hair in her hand and began to twist it into the form of a horn. Gradually she was transformed into a narwhal. They swam away together and to this day can be seen swimming side by side in the waters of the Arctic Ocean. The narwhal and the man in the sculpture below represent a variation on the legend.*

Chapter II. The Challenge

Civilization first appeared in the valleys of ancient rivers. The rivers provided man not only with a wealth of water for his crops and an abundance of fish, but also with a safe and easy path up and down the land for the exchange of ideas, artifacts, and people, all of which helped promote the growth of civilization. Since all these rivers led to the sea, man was eventually tempted to venture out onto a dangerous ocean.

The form of the first boats is a matter of conjecture. The first vessels may have been the outriggers of the Pacific, the skin boats of the Eskimos, or the rafts of the American Indians. Whenever they were out of sight of a familiar point of land, they steered by the stars. Some sailors on the Indian Ocean had a means of navigation that was available only to them and only at certain times of the year—they were guided by the monsoons. Some people did not take to the sea by choice, but were forced to do so. Probably the first of such people were the Polynesians, who were compelled to make a hazardous journey by sea to new lands. About the third century B.C. they were driven from their homes on the Asian mainland and began moving eastward across the Pacific from island to island. They are supposed to have enjoyed yet another novel means of navigation—they followed the migrations of birds. Roughly a thousand years later, their migrations had taken them as far as the central Pacific. They made the journeys in great double canoes. As many as 300 of these canoes traveled together.

"Now they made all secure in the fast black ship, and setting out the winebowls all abrim, they made libation to the gods, the undying, the ever-new, most of all to the gray-eyed daughter of Zeus. And the prow sheared through the night into the dawn."

Homer, *The Odyssey*

These Polynesians and all the other brave people who challenged the sea had to face not only real but also legendary hazards. The boats of the early challengers were not very seaworthy and this made the danger very real; but they also had to overcome their imagination. Early man peopled the sea with monstrous and terrifying beings. Only by meeting the challenge could these myths be dispelled as they proved to be imaginary.

Not only the early voyagers but the curious men who stayed on land began to question some of the myths of the sea and to make observations for themselves. In this way, natural history—an attempt to give a more rational explanation to the phenomena of the universe—began to grow. Because the seas beneath the surface could not be observed, the natural historians innocently compounded some of the myths. One of the greatest of these early natural historians believed, for example, that all the seas were purified at each full moon. It was undoubtedly the fact that a greater quantity of matter is cast on shore at this period that led him to this conclusion. And all the time that old answers were being questioned, new ones were being created in man's imagination.

One-eyed giants. *Odysseus and his men came to the land of the Cyclopes and blinded their leader Polyphemus in order to make their escape. The giants hurled boulders at the fleeing ships as depicted in this section from a fresco in the Vatican Library.*

The Long Way Home

One of the most harrowing legends pitting men against the sea is Homer's epic the *Odyssey,* telling of the wily Odysseus and his struggle to return home to Ithaca after the Greek victory over Troy.

Odysseus had been reluctant to leave his wife and infant son to go to war; it had been predicted that he would be away for 20 years. The war had lasted a full ten of them. At the end the Greeks' powerful champion, Poseidon, had ominously turned against the victors.

As they set out, a gale came up and blew for nine days, sending the ships to the land of the Lotus-Eaters. There some of the men ate a sweet fruit and lost their desire to continue the journey; they had to be dragged back to the ships. In Sicily the giant one-eyed Cyclops, Polyphemus, son of Poseidon, took Odysseus and a dozen men captive, secured

***Odysseus** had many dangers to overcome during the ten years that he sought to return to his home in Ithaca after the Trojan War. One of the most fearsome was the alluring song of the Sirens, who tempted sailors to come to them and thereby give up their freedom. When Odysseus and his men passed the land where they dwelled, he had the crew stuff wax in their ears, but he had himself tied to the mast so that he might hear their song. This third-century mosaic from Tunisia (above) depicts that scene.*

them in his cave, then began to eat them, two at a time. After six had been devoured, Odysseus got the giant drunk on wine and blinded him as he slept. When the wounded Polyphemus moved the boulder at the cave entrance to let his flock out, the men escaped by concealing themselves beneath the sheep. The episode further infuriated Poseidon.

The travelers next came to the island of kindly Aeolus, keeper of the winds, who gave Odysseus a leather bag containing all the winds but the westerly, which would blow them home in nine days. While Odysseus was asleep, his men opened the bag and the freed winds blew them back to Aeolus, who was angered by their return and ordered them away. Next, 11 of their ships were shattered by cannibals. After that they dallied for a year with Circe, a sorceress who was able to turn men to swine. They sailed past the Sirens and avoided the dread whirlpool Charybdis, but lost six men to Scylla, a six-headed monster. Hungry, they killed the forbidden cattle of Helios at their next stop. For this they were punished with a terrible storm, which killed everyone except Odysseus. Odysseus fetched up on Calypso's island where he spent seven years, yearning for home. Finally released by the nymph, he built a raft and sailed away, only to be wrecked on the island of the Phaeacians, who took Odysseus home at last. For doing so, Poseidon turned their ship to stone.

Jason and the Argonauts

When Jason was 20, he came to Iolcus in Thessaly to claim its throne from his uncle, the usurper Pelias. Pelias told his nephew he could have the throne if he would first travel to Colchis and bring back the Golden Fleece guarded there by a sleepless dragon. The venture appealed to the young man; he accepted the challenge and so began the epic sea voyage of Jason and the Argonauts.

The goddess Athena personally directed the construction of their ship, seeing that its stout pine planks were bolted and fastened with a tight girdle of well-twisted rope to withstand the battering of the waves. *Argo,* as it was named, had benches for 50 rowers and a mast for sail. The crew under Jason's command was made up of 50 heroes from all over Greece. After making sacrifice and praying to Apollo for a safe return, they feasted, slept, and then boarded ship.

Their first landfall was the island of Lemnos where they dallied with a feminine population that had killed its menfolk the year before. Only the wrath of Heracles brought the Argonauts back to their oars. In dark of night they hugged the shore opposite Troy and slipped past their enemy King Laomedon and into the Sea of Marmara. They encountered and defeated the fierce six-handed earthborn giants; then farther on one of the crew felled King Amycus in a boxing match.

At Salmydessus they aided blind Phineus by ridding him of a pair of loathsome Harpies. He repaid them with wisdom. Following his advice, they freed a dove as they approached two floating rocks guarding the Bosporus. The rocks clashed together as the dove flew between them, nipped off some tail feathers, then recoiled. At that moment the Argonauts stroked hard and got through before the rocks came together again. They lost only a stern ornament. At an island sacred to Ares they wore protective helmets against the dive-bombing birds that Phineus had warned them of, and they shouted and beat their shields with swords to scare them off.

Aeëtes, king of Colchis, agreed to give up the Golden Fleece only if Jason would perform impossible feats. He was able to do them with the help of the king's daughter Medea, who had fallen deeply in love with him. But in the end they had to steal the prize. Medea sailed away on *Argo* while Aeëtes followed angrily in pursuit. It is said that Medea cut up the body of her little brother and threw the pieces in the sea so that Aeëtes would be delayed when picking them up.

Argo sailed up the Danube, then to Libya. For 12 days they carried the ship over land. Like Odysseus, who would later make a similar voyage, they ran into Circe. They were able to resist the Sirens because crewman Orpheus sang songs even more enticing than those of the Sirens. The gods helped them get safely by Scylla and Charybdis. At the island of the Phaeacians Jason and Medea were married.

When they got back to Thessaly, Jason gave Pelias the Golden Fleece. But the crafty old man revoked his promise: Jason would never occupy the throne at Iolcus. Medea squared matters in her usual way by tricking his daughters into killing Pelias; after that the newlyweds wisely found refuge in Corinth, where they beached *Argo* and consecrated it to Poseidon. One day when Jason was old, he crawled beneath the ancient hulk to rest. The stern broke off, fell, and killed him.

The return of the Argonauts *is fancifully depicted by a Renaissance painter who advanced shipbuilding and costume design many centuries beyond that of Jason's period in history.*

Cargo for Rome. *Two merchant vessels approach the Portus lighthouse at Ostia, a harbor built at the mouth of the Tiber to serve Rome. The ships' sails were made stronger by strips of leather.*

The King Who Watched a Sea Battle

In 480 B.C. Xerxes, the king of Persia, sat on a hill in Greece and watched one of the greatest naval battles of all time—Salamis. But the king did not enjoy what he saw.

Three hundred and seventy-five Greek ships with 80,000 aboard met more than a thousand Persian triremes manned by 120,000 men. The outcome of that battle played no small part in determining the future of western civilization.

Xerxes thought that he had superiority not only in numbers, but also in morale. He was wrong because he was fighting against men who were defending their homeland. Themistocles commanded the Greeks, and naval historians say that his plan of battle was the first in which a fleet prepared to defeat an attacker by taking up a position that flanked his advance. He drew his triremes up in a column, 15 ships in line and 25 deep with 100 yards between them. As the Persian fleet came out of the straits, the Greeks came on, their oarsmen straining every muscle to attack at full speed. Before the Persians were able to form a line of defense, the Greek ships rammed them and their warriors boarded them. Many of the Persian triremes were destroyed in the engagement, and the others, although they succeeded in getting

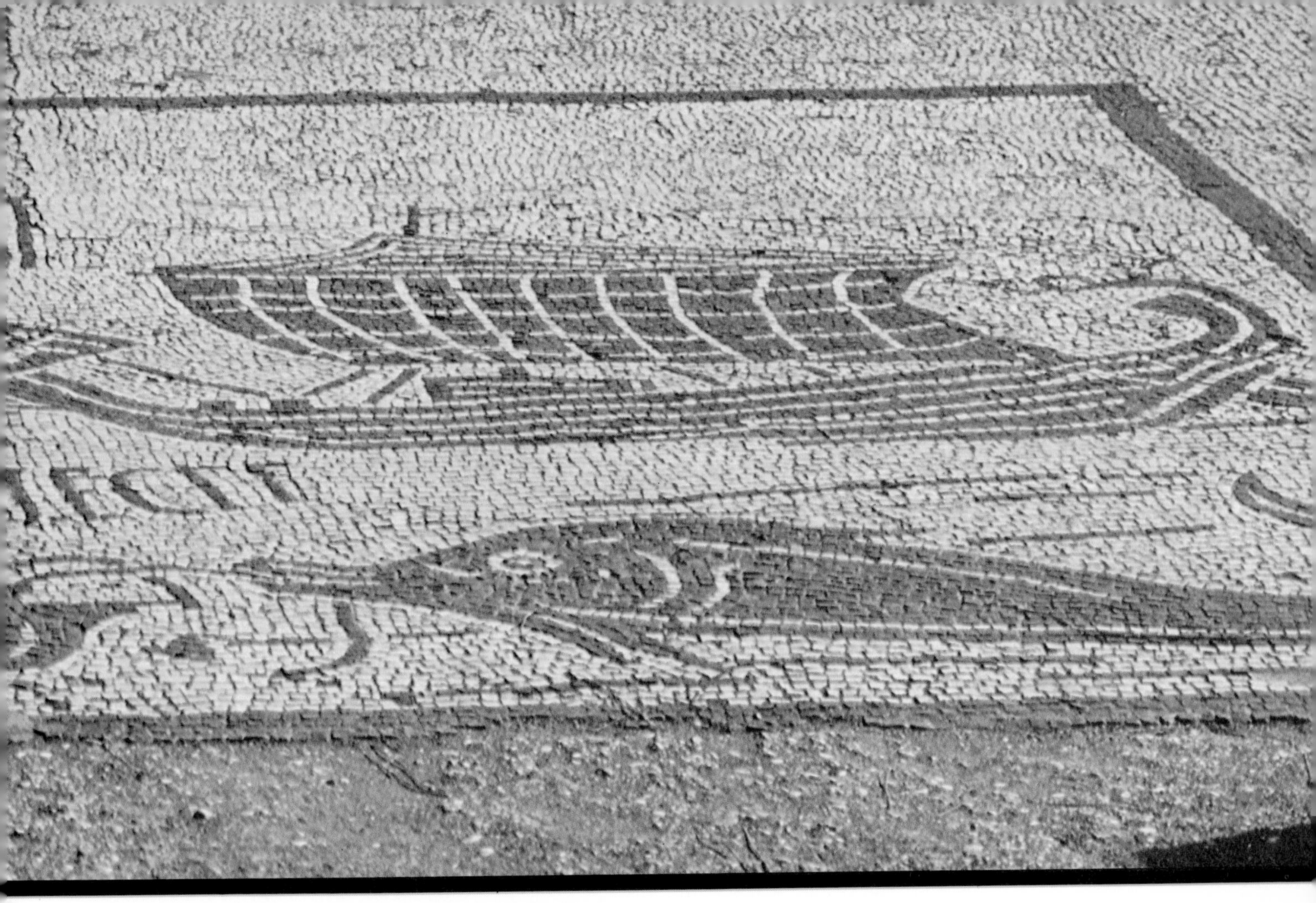

Sidewalk mosaic *(above and right), from the second or third century A.D., is a section of an advertisement for the nearby offices of an Ostian shipping company, Naviculari Syllectini.*

into the bay, never managed to recover a solid formation. Few men in the Persian fleet could swim, and those who were forced over the side to avoid the sword were drowned. The Greeks, on the other hand, were just as much at home in the water as they were on the land. Those who lost their ships threw away their armor and their weapons and swam to shore or to another ship where they joined the fight again.

Soon the Persians were in flight, and Xerxes knew that the invasion of Greece was over, for he needed control of the sea to support his army. The battle opened the way for the dawning of the Golden Age of Greece.

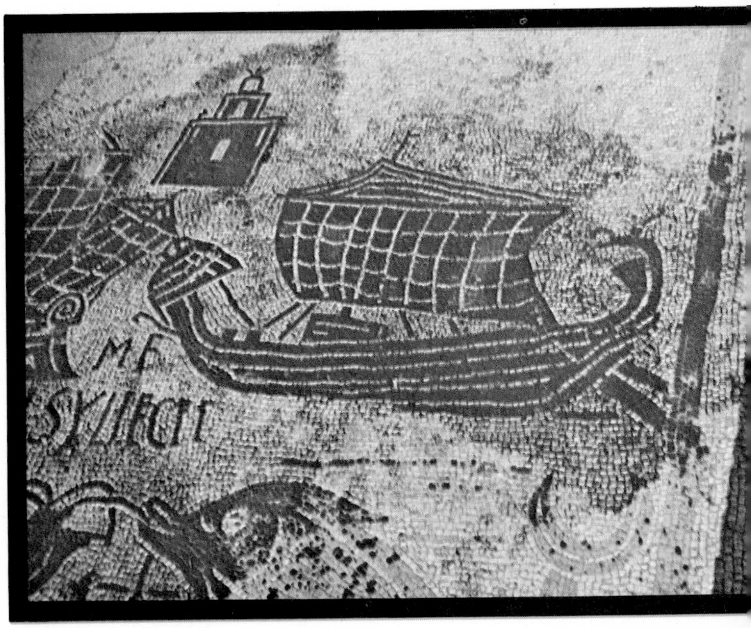

The Infancy of Marine Biology

One of the most industrious of scientists was Pliny the Elder, who lived in Rome in the first century A.D. His great work was a 37-volume encyclopedia of natural science called *Historia naturalis,* which dealt with the whole physical universe. Most of Pliny's information came to him secondhand and his work is quite useless as science, but this does not take away from his achievement, for he had made a beginning. Others would follow along the path.

Pliny the Elder devoted Book IX to fish, and what he wrote reveals how much natural science was intermingled with tales and imagination: "All fish have a presentiment of a rigorous winter, but more especially those which are supposed to have a stone in the head, the lupus, for instance, the chromis,

the sciaena, and the phagrus. When the winter has been very severe, many fish are taken in a state of blindness. Hence it is, that during these months they lie concealed in holes, in the same manner as land animals, as we have already mentioned; and more especially the hippurus and the coracinus, which are never taken during the winter, except only on a few stated days, which are always the same. The same with the muraena also, and the orphus, the conger, the perch, and all the rockfish. It is said that, during the winter, the torpedo, the psetta, and the sole conceal themselves in the earth, or rather, I should say, in excavations made by them at the bottom of the sea."

Pliny's remarks about love and hate among the fishes are amazingly funny for a contemporary skin diver to read: "The mullet and the wolffish are animated with a mutual hatred; and so too, the conger and the murena gnaw each other's tails. The crayfish has so great a dread of the polypus, that if it sees it near, it expires in an instant: the conger dreads the crayfish; while, again, the conger tears the body of the polypus. Nigidius informs us that the wolffish gnaws the tail of the mullet, and yet that, during certain months, they are on terms of friendship; all those, however, which thus lose their tails, survive their misfortune. On the other hand, in addition to those which we have already mentioned as going in company together, an instance of friendship is found in the balaena and the musculus, for, as the eyebrows of the former are very heavy, they sometimes fall over its eyes, and quite close them by their ponderousness, upon which the musculus swims before and points out the shallow places which are likely to prove inconvenient to its vast bulk, thus serving it in the stead of eyes."

All the seas are purified at the full moon; some also at stated periods. At Messina and Mylae a refuse matter, like dung, is cast up on the shore, whence originated the story of the oxen of the Sun having had their stable at that place. To what has been said above (not to omit anything with which I am acquainted) Aristotle adds that no animal dies except when the tide is ebbing. The observation has been often made on the ocean of Gaul; but it has only been found true with respect to man.

Pliny (c. A.D. 23–A.D. 79)
Natural History, *Book II*

Chapter III. Golconda

In the state of Hyderabad in India there is a city called Golconda—the City of Diamonds. In the Middle Ages it was only one of the many places in the so-called Indies where great wealth was to be found. Marco Polo had brought back stories of fabulous wealth and so had later travelers by sea and land over an eastern route from Venice and Genoa and other great commercial cities of the late Middle Ages. Not only were the Indies incredibly rich in precious stones but in much less glamorous but no less valuable commodities such as spices. The drive to acquire this wealth became the same as the drive for power.

> "Ah! If I had only been master of the seas."
> Napoleon, in exile on St. Helena

In 1626 a subordinate of Cardinal Richelieu wrote that "whoever commands the sea will have great power on the land. Look at the King of Spain. After he had mastered the seas he was able to conquer so many kingdoms that the sun will never set on his possessions!" First Portugal and then Spain had shown that this was indeed so. They divided half the oceans of the world between them. But when the Dutch shattered the declining naval power of Spain at the Battle of the Dunes in 1639, the struggle on the seas became a confusing, many-sided contest among the great powers of Europe. Throughout the seventeenth and eighteenth centuries there was almost continual fighting on the ocean. The Dutch defeated the English in the Four Days' Battle in 1666 and for a time had the upper hand. Toward the end of the eighteenth century two great powers had emerged—France on land and England on the sea. The outcome of that rivalry was determined on the sea, and England won it, first by defeating the French at Quiberon Bay in 1759, then again on the "Glorious First of June" in 1794, and, finally, resoundingly, at Trafalgar in 1805.

As populations increased, the need for fish grew. By the eighteenth century all the great maritime nations of Europe were fishing the waters of the world on a large scale. Holland's prosperity was built on its fisheries. Of the many types of fishing, whaling was the most profitable. The first whalers were the Basques, but by the eighteenth century England and Holland dominated whaling. At the end of that century the Americans had become very active. Soon all the great seafaring nations had joined in the hunt of the whale, and it was not long before populations were seriously depleted in some parts of the world.

From ancient times to the present day, the greed to possess the wealth of the sea and the wealth that the sea leads to has triggered devastating conflicts. Today the situation is as bad as it has ever been. Nations still claim exclusive rights to ocean domains and those that have the technological means fish the oceans rapaciously. There are frequent confrontations between ships of two nations claiming fishing rights or jurisdiction. These claims and counterclaims must be resolved before they are allowed to lead to a conflict that could have devastating consequences.

Methought I saw a thousand fearful wrecks,
Ten thousand men, that fishes gnawed upon;
Wedges of gold, great anchors, heaps of pearl,
Inestimable stones, unvalued jewels,
All scattered in the bottom of the Sea.

Shakespeare, Richard III, *I, iv*

Manna of the Mediterranean

Before setting sail or after returning safe from the sea, fishermen of ancient Greece offered gifts to the gods. Poseidon came first in their veneration and after him Hermes, or Mercury. Pan, a son of Mercury, and Priapus, the god of propagation and fecundity, were also revered and honored by the fisherfolk.

This was only after fish became acceptable as food. For a long time the only fish that the early Greeks ate were the marvelous eels of Lake Copaïs. Neither fishing nor fishermen have any status in the *Iliad* or the *Odyssey*. No fisherman is regarded as a contributor to the wealth of the tribe or state.

This is understandable. The ancient Greeks, like all early peoples, disliked long sea voyages. They shrank from spending the night on the water. They would go three times the distance if only they could keep the land always in sight. Besides, the Aegean was only safe for their ships a few months out of the year. There is in the *Odyssey*, however, mention of a "blameless king" as one who rules where "the black earth bears wheat and barley, and trees are laden with fruit, and sheep bring forth and fail not, and the sea gives stores of fish." Plato, who conceived of an ideal state, was not fond of fishing: "O Friends," he said, "may no desire of hunting in the sea, or of catching the creatures in the waters, ever take possession of you."

As the population in the Mediterranean lands grew, the people began to turn more and more to the manna of the Mediterranean to feed them, but for a long time fish was to remain the food of the poor.

Ocean harvest. *In some parts of the world the people give thanks to the sea for the food it provides and propitiate the fish in hope of an abundant catch. In Spain there are festivals like the one shown here as drawn by a child of that country.*

GEMA

The First Poachers

It was not long after people began to fish that they began to argue about who the fish in the sea belonged to. The earliest record that we have of poaching—stealing another man's fish—is in a letter written by the successor to Hammurabi, the great Babylonian king who drew up a famous code of laws. That code of laws sought to make clear what individual property meant. Now Hammurabi's successor, Sansu-iluna, expanded that concept. "They have reported unto me," he wrote, "that the ships of the fishermen go down into the district of Rabim and to the district of Shakanim and catch fish. I am therefore sending unto thee an official of the palace gate. When he shall reach thee, the ships of the fishermen which are in the district of Shakanim shalt thou recall, and thou shalt not again send the ships of the fishermen down into the district of Rabim or the district of Shakanim."

Even then people maintained that they had exclusive rights to fish in their own home waters. But whether or not this was fair to the others was a question that has not been resolved to this day.

Today, claims of sovereignty are extended further than ever before—in many instances to 200 miles from shore. Obviously this is not fair to the twenty-nine landlocked nations. Perhaps a solution will be found in dividing fishing rights in the oceans among all the nations of the world irrespective of contiguity. We would have to know a great deal more about population dynamics in the sea than we do now, however, if such a plan were to be successful.

Fishermen, *from the temple floor of Piazza Armeria in Sicily, take their catch with a net, spear them with a trident, and hook them on a line.*

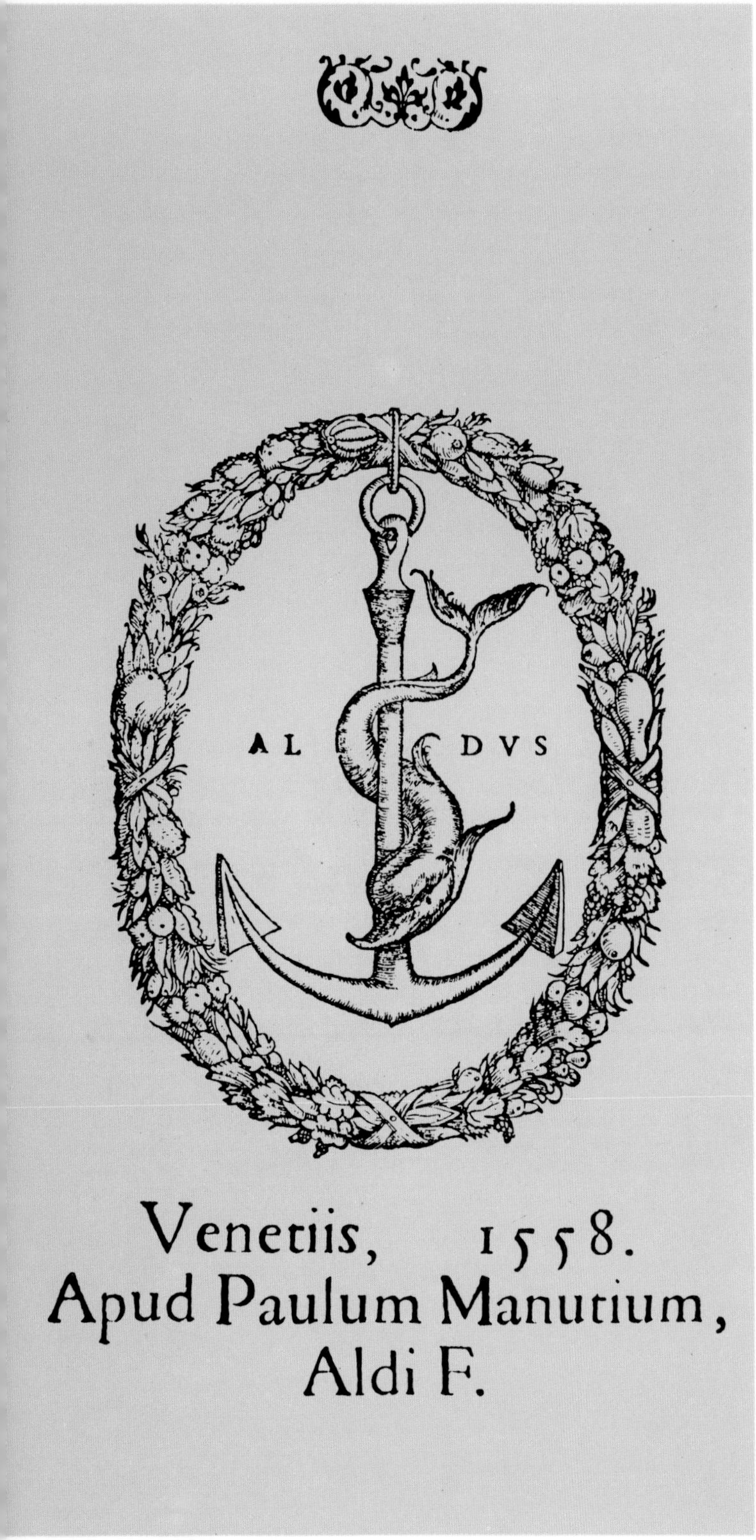

Images from the sea have found countless uses. Postage stamps have been designed bearing pictures of fish and other sea motifs. In the press mark above, a dolphin is wrapped around an anchor, a symbol for the motto ***Make Haste Slowly.***

Some Fantastic Islands

Some legendary creations of navigators were passed on to the cartographers.

At the time of the conquest of Spain by the Moors, it was said, seven bishops and a great number of their followers took flight in ships that could not sail and abandoned their fate to the seas. After tossing about for many weeks, they landed at an unknown island somewhere in the Atlantic. The bishops burned the ships to prevent their flock deserting and founded seven fabulous cities. The Island of the Seven Cities was identified with the one mentioned by Aristotle as having been discovered by the Carthaginians and was put down in early maps about the time of Columbus as the island of Antilla.

The Irish have always been intrigued by islands and have given the world tales of an island of monstrous ants, an island of bloodthirsty quadrupeds, the island that died, the island of the burning river, the island of weeping, the isle of speaking birds, and an island that stands on a pillar.

There have been countless tales of islands that disappeared and then were seen again, and of islands that seemed to float away. Undoubtedly these were in fact small islets that the tide submerged and uncovered.

A cartographer named Bianco projected a map in 1436 that had on it the Island of the Hand of Satan. It was believed by some that there was an island in the shape of a hand that only made its appearance in foggy weather or at night, and unlike neighboring islands, it was always cloaked in chilly mists.

Before the invention of photography, most of our visual knowledge of the sea came to us from the ***drawings of the naturalists*** *(right), some of whom were extremely talented artists. Most of them interested themselves in all forms of life.*

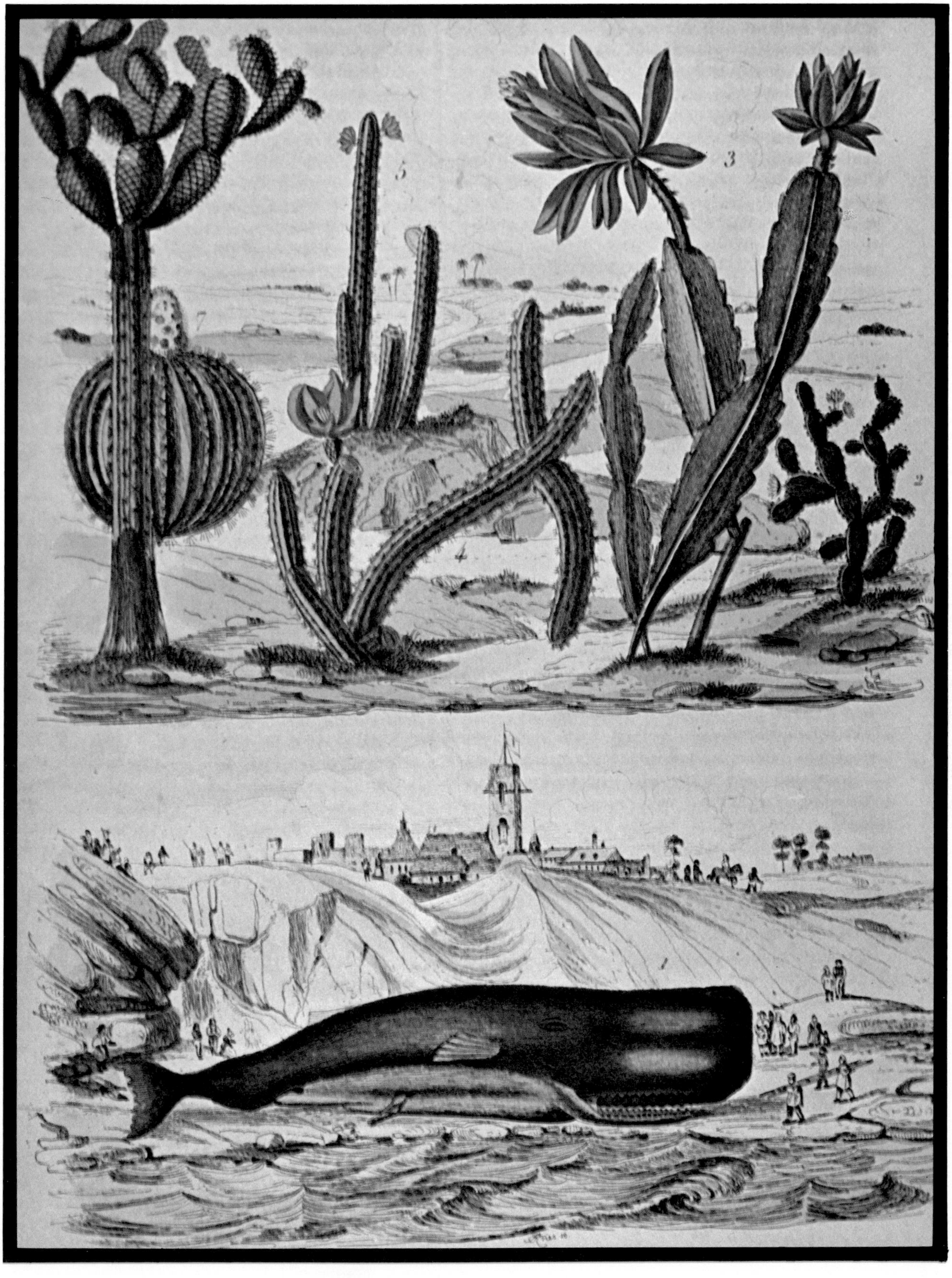

The Welcome Light

As the Mediterranean became an avenue of commerce, there were naturally an ever increasing number of wrecks at sea. To avert disasters near their own harbors, Ptolemy I and Ptolemy II, pharaohs of Egypt, built a great lighthouse at Alexandria, completing it in about 279 B.C. Its beacon could be seen for 25 miles at sea and must have saved many a ship bound for that port. It probably stood until the fourteenth century, when a great earthquake sent it tumbling to the ground. As long as it remained, it was known as one of the seven wonders of the world.

No less remarkable was the Colossus of Rhodes, a huge hollow bronze statue of the sun god, Helios; according to some accounts, it held a burning torch aloft. It was completed in about 280 B.C.; it is said to have been more than 100 feet high and to have straddled the entrance to the harbor of Rhodes so that ships passed between its legs. It, too, was destroyed by earthquake, crumbling in 224 B.C. Until 656 A.D. its ruins lay undisturbed; then a Moslem trader bought the metal to melt into cannon.

The Romans were the greatest builders of lighthouses in the ancient Mediterranean. They built one at Caepio in Spain in 20 B.C. and at Ostia, the port of Rome, some 75 years later. Then they went on to construct no fewer than 30 more, scattered here and there for the protection of the ships of their expanding empire.

The most unusual lighthouse, if it can be called that, was one said to have been the invention of Charlemagne. He ordered that trumpeters were to be stationed at a chapel at the mouth of the Gironde River to warn ships off the rocks when the fog was heavy.

Today lighthouses have passed their usefulness in some parts of the world, while in others they remain just as welcome a light as they were almost 2000 years ago.

Navigational aides. *Lighthouses marking dangers to shipping have served since pre-Christian times. Today's satellites are tomorrow's navigational tools.*

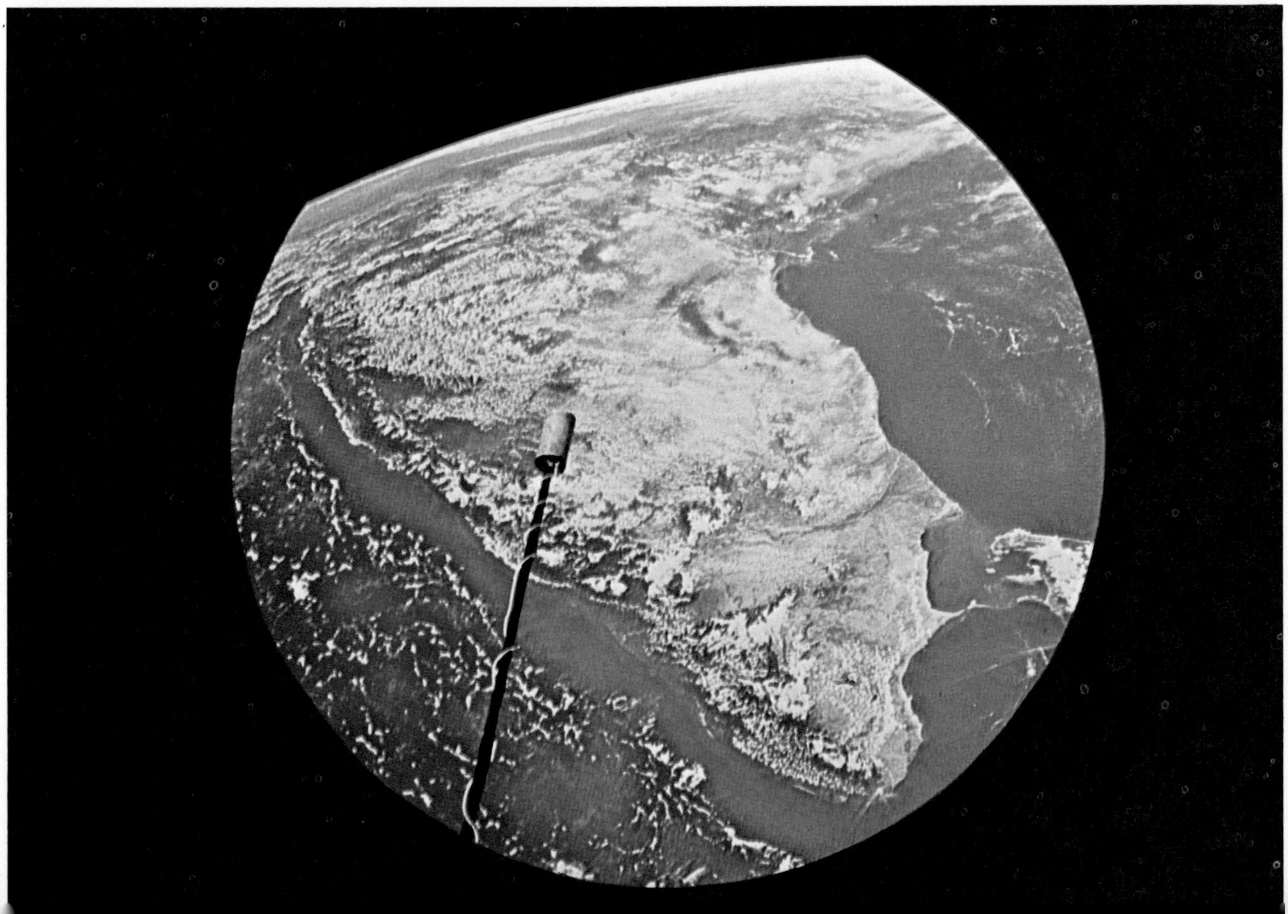

Struggle for supremacy on the sea *was constant in the sixteenth and seventeenth centuries. In this painting by Hogenbergh (above) the forces of Spain and France clash in 1573. Spanish sea power was destroyed by the English 15 years later.*

A man of war *under full sail salutes in this oil on canvas by the Dutch painter William van de Velde (1633-1707). Van de Velde's specialty was naval battle scenes. Each detail of the painting (right) is done with perfect accuracy.*

The Fate of the Invincible Armada

The sea demands respect; it treats those who belittle its power to unforgettable lessons. Such a lesson was in store for the Spanish fleet, which called itself invincible, when it set out to conquer the English in 1588.

The English had been raiding King Philip's realm in the New World and had been capturing his treasure ships on the high seas. The Spanish king therefore built a great fleet, the Armada, to conquer England and to establish on the island a government he could control. When the Armada set sail it consisted of 128 ships and 30,000 men.

The English ships that met the Spanish were largely raiders and merchantmen. They were fast ships and quite maneuverable compared to the Spanish vessels, which were clumsy and incapable of sailing close into the wind. The English engaged the Spanish fleet with a tactic never before employed in naval warfare. Instead of fighting the enemy at close quarters, their ships dodged about,

firing at an enemy ship itself rather than at the men aboard her. This was certainly the correct use of cannons, but either no one had thought of it before or they did not have the ships capable of it. For ten long summer days the fleets clashed, and finally the Spaniards were defeated. But it was not the English who defeated them—it was the sea. In the course of the battle a great storm arose and drove the Armada blindly east-northeast up the coast. By the ninth day the Spanish fleet was scattered over the North Sea with the English in pursuit. On the tenth day two Spanish ships were sunk. They were the first and the last destroyed on either side by direct enemy action. Yet in all, the Spanish fleet lost 54 ships and most of its men—all to the sea, which drowned them and battered them to death on the rocks. Spain's flamboyant bid to govern the sea had come to an ignominious end.

England remained free to become the greatest naval power the world had seen—and thus, for a long, long time, to control the seas.

Every Man Will Do His Duty

At the western entrance to the Strait of Gibraltar there is a low cape on the Spanish side. Off this cape in 1805 a naval commander with one eye and one arm routed the enemy at sea and so saved his country from invasion by a man who threatened to conquer all of Europe. That cape was Trafalgar, and the commander was Lord Nelson.

Nelson had drawn up a plan of action to meet the combined fleets of Napoleon and the Spanish. He did not want an orderly contest of lines of ships maneuvering around each other but a free-for-all in which all ships would be at close quarters. Nelson's second-in-command would break the enemy's line in front of its last 12 ships and destroy these. Nelson himself, with most of the remainder of the fleet, would break the enemy formation closer to its center and prevent any attempt to rescue the 12 that were cut off.

When the two fleets met, the French commander had his ships in a curved line perpendicular to the oncoming English vessels. Thus the French could deliver far more firepower broadside than the English from their bows. This did not discourage the English. They knew they were better sailors and better gunners than the enemy and they knew that the French made the great mistake of firing at the masts and rigging instead of the hull. Just before the first shot was fired, Nelson gave the signal: "England expects that every man will do his duty."

And they did it very well. Napoleon's dream of conquering England had ended in a nightmare. In England, the joy at the news was muffled by sadness. Nelson was killed leading the fight.

A sixteenth-century view of ***the harbor of Naples*** *by Pieter Bruegel, the elder.*

Gilgamesh

One of the oldest and most enduring stories in the world is that of Gilgamesh, who loved his friend and lost him to Death and then discovered that he lacked the power to bring him back to life.

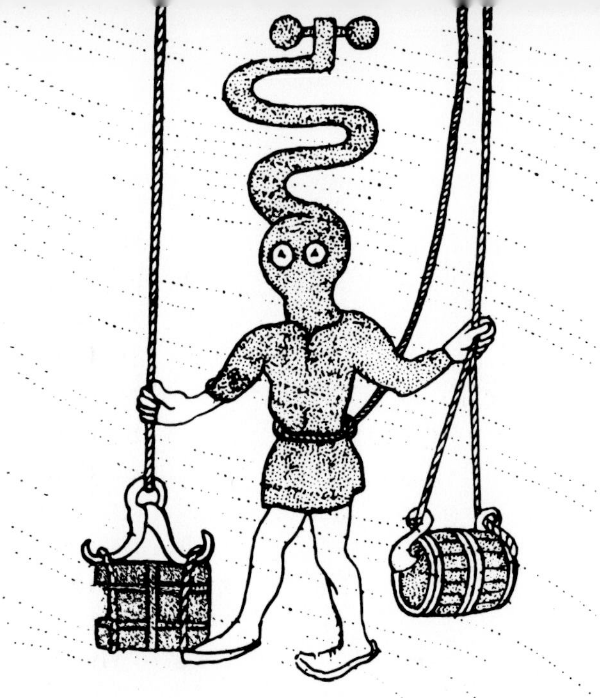

The epic was Sumerian in origin and was later added to and unified as a national epic by the Semitic Babylonians, who had succeeded to the earlier culture and civilization of the valley of the Tigris and the Euphrates. It is older than either the Bible or the works of Homer, predating the latter by at least a millennium and a half. Composed in some 3000 lines, it was inscribed on 12 stone tablets discovered much later among the ruins of the city of Nineveh. It was probably known to early biblical authors. The epic contains a story of a flood that is similar in many ways to the story of the flood in Genesis. The boat of the earlier flood story carries one family and animals but also a navigator and the craftsmen of the city. The raven and the dove of the Noah story are joined by a swallow.

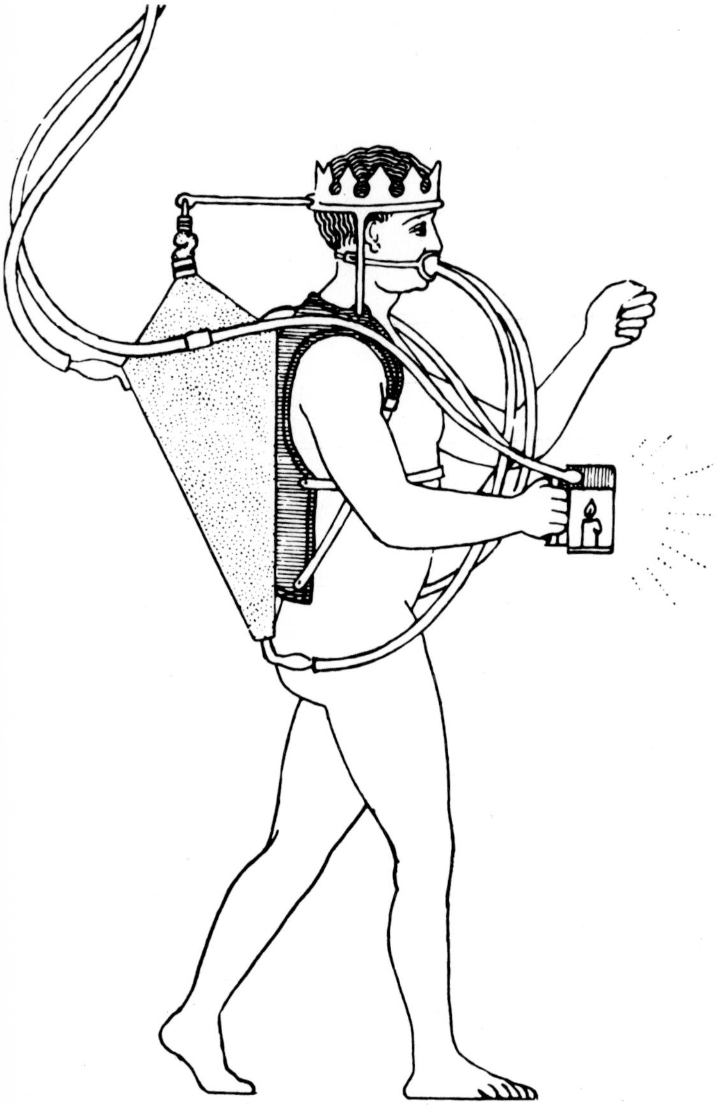

Gilgamesh was a bellicose and imperious king of legend. The epic tells of his adventures with his friend Enkidu, and how both become more human through their friendship. When Enkidu suddenly falls ill and dies, Gilgamesh, heartbroken, sets out to find the secret of eternal life in order to bring his friend back to life.

Gilgamesh makes the dangerous journey over the Sea of Death to Utnapishtim, the wise man, whose name means "He who saw

*The drawing of a man (above) wearing a **leather helmet** with a long breathing tube dates from the fifteenth century. Diver (left) wears Drieberg's **Triton,** designed in 1808.*

life." Utnapishtim and his wife were the only survivors of the great flood that destroyed every living thing on earth. The wise man tells Gilgamesh of a plant at the bottom of the river that will give him new life. Gilgamesh goes to the river, ties stones to his feet, and dives into the water. Searching the bottom, he finds the plant. Rose-colored and ambrosial, it shimmers in the river like a prism held to the sunlight. When he grasps it, it cuts into his palms, and he watches his blood flow into the water. Then he cuts the stones loose from his feet, rises to the surface, and swims to shore shouting, "I have it! I have it!" A boatman takes him to the opposite bank of the river where he pauses to rest and refresh himself, leaving the plant unguarded. A serpent smells its sweet fragrance and comes out of the water and devours it. When it does, it sheds its skin. The serpent has found the secret to renew its life. Gilgamesh had that secret and lost it.

The earliest known imaginary ***diving device*** *is represented in this Assyrian bas-relief from about 900 B.C. in Ashurnasirpal's palace. It shows diving warriors crossing a river.*

A Treasure Map

If you draw a line on a map from Colombo in Ceylon to Freemantle in Australia and put an "X" on the map about 12° south of the equator you will have a treasure map. For it is at this point, somewhere on or in the vicinity of the Cocos or Keeling islands, legend says, one of the most fabulous pirate treasures of all time is buried.

The pirate was Captain Edward Davis, and he was indeed a very unusual pirate. He never killed anyone, the way Captain Kidd or some of the other notorious buccaneers did. He merely stole from them. And when he got tired of being a pirate, he paid the king a bribe, and the king then pardoned him for all his past misdeeds. Then he retired from pirating and settled down to the life of a respectable gentleman. Perhaps he had enough transportable treasure with him that he never felt compelled to return to the Indian Ocean to reclaim the wealth he had left there. And there it is to this day, so far as we know. Hundreds of salvagers have sought for it, but with no luck.

Probably the greatest cache of sunken treasure is scattered along the Spanish Main and along the route from there to Iberia. The Spaniards were among the greediest, and they took all the gold and jewels they could find in the New World and shipped it back to the Old. But the English and others were no less greedy, and they raided those ships for their cargoes, and many of them were lost. They are still there, and others will find them some day.

***Malaysia** has one of the longest coastlines in the world relative to its area. It is not surprising, then, that the sea and ships are favorite images in the drawings of children of that nation.*

Mare liberum

In 1604 the Dutch East India Company, in order to justify the action of one of its admirals in seizing a Portuguese vessel, asked the jurist and statesman Hugo Grotius to write a treatise on the lawfulness of the capture of merchant ships.

Grotius did much more than that. He began by describing the law of mankind in general. He drew on poets as well as philosophers of law for his argument, citing Hesiod, Cicero, Horace, and Vergil among others. As he went on, he went even further back to first causes, describing the law of nature as it derived from God's will. He concluded that, because Portugal deprived the Dutch of the right to sail to the East Indies for commerce, the East India Company was entitled to capture Portuguese merchantmen to compensate itself for its losses. Grotius maintained that the ocean is free to all nations. His decision was published in 1609 as *Mare liberum* (freedom of the seas).

Grotius closed his argument with a dramatic plea to regard the ocean in the way it had

once been considered: "The question at issue then is not one that concerns an inner sea, one which is surrounded on all sides by the land and at some places does not even exceed a river in breadth, although it is well known that the Roman jurists cited such an inner sea in their famous opinions condemning private avarice. No! The question at issue is the outer sea, the ocean, that expanse of water which antiquity describes as the immense, the infinite, bounded only by the heavens, parent of all things; the ocean which the ancients believed was perpetually supplied with water not only by fountains rivers, and seas, but by the clouds and by the very stars of heaven themselves; the ocean which, although surrounding this earth, the home of the human race, with the ebb and flow of its tides, can be neither seized nor inclosed; nay, which rather possesses the earth than is by it possessed."

Therefore the sea can in no way become the private property of anyone, because nature not only allows but enjoins its common use.

Grotius, Mare liberum

Man's Past Beneath the Sea

Nearly everything that man has ever made has been carried, at one time or another, in ships, and many ships have sunk. Scattered about the bottom of the sea are some of the richest archaeological treasures in the world.

Most sites are found by accident, through chance discovery by fishermen, sports divers, and spongers. A few, however, have been determined by scientific research in documents relating to the sailing of a particular ship. The most spectacular of such finds, although it is too recent to belong to archaeology, has been that of the *Vasa,* a seventeenth-century Swedish warship that was raised from the harbor of Stockholm in 1961. The *Vasa* was relatively well preserved because the low salt content of the Baltic Sea does not support the life of the teredo, the shipworm that rapidly destroys wooden remains in most seas and oceans. Since then, divers have continued to comb the area for items necessary to a total restoration of the ship, and they have already brought up more than 1000 pieces, including more than 200 sculptures and other ornaments. Today the *Vasa* is housed in her own museum.

These are among the countless artifacts and works of art of inestimable value that have

*The nymph **Arethusa** surrounded by four dolphins is engraved on a 479 B.C. commemorative silver coin of Syracuse (below). Arethusa dived into the sea near Olympia to escape the ardor of the river god Alpheius. When she surfaced in the harbor of Syracuse, she came under the protection of the goddess Artemis, who transformed her into a freshwater spring which bubbles to this day.*

A plate from the fourth century B.C. with the image of a fish. In ancient Syria fish were considered sacred. The Greeks were impressed by that reverance, and the early use of fish as a Christian symbol may be in part associated with its sanctity in Syria.

been raised from many sites all over the world but especially in the Mediterranean, where more archaeological work has been done than anywhere else.

Archaeology has taught us some interesting things about some of the ancient sea routes, as well. For example, a cargo of bronze objects was dredged from waters off the coast of Spain. They appear to have been lost in the seventh century B.C. The swords are similar to those found in Sardinia. The finding gave evidence of a possible trade route between the Atlantic and the Mediterranean almost 3000 years ago!

The best prospect for efficient archaeological work undersea lies in the development of our capacity to live underwater for extended periods of time. In the Conshelf II village, which we established in the summer of 1963 beneath the Red Sea, six men lived for a month in a house 36 feet deep, with two men staying in a deep cabin at 90 feet for a week. Since then, we have had men living at 330 feet and working at a depth of 370 feet in the Mediterranean off the south of France. The day is not far off when archaeologists will live in such houses, and excavations that might have taken many summers to complete will be done in one.

Chapter IV. Elsewhere

Seafaring in the west had developed in and around the Mediterranean. The inland seas had been challenged, and some had even glimpsed the great oceans that lay beyond. In the sixth century B.C., Hanno of Carthage passed through the Strait of Gibraltar and visited the west coast of Africa. In the fourth century B.C., Pytheas went from Marseilles around England and sailed as far north as Iceland.

The Norsemen were among the first explorers to give vent to their curiosity, and the desire for the wealth that lay "elsewhere." The poverty of the soil had forced them to turn to the sea for food, and it had bred in them great courage and bold seamanship. Their voyages were to take them from Norway to the Mediterranean as far as Constantinople and, more significantly, to the Faroes, Iceland, Greenland, and finally, America.

"Let our ocean-striding ship
Explore the broad
tracts of the sea
While these eager swordsmen
Who laud these lands
Settle in Furdustrands
And boil up whales."

Eirik's Saga

One of their first explorers was Ottar, and he was one of the few who went simply out of curiosity. He sailed around the North Cape and as far as the White Sea because, as he said, he was "desirous to see how far the country extended north and whether anyone lived there."

By the fifteenth and sixteenth centuries, all the seafaring nations of Europe were ready to join in what was to be called the Great Age of Discovery. Great strides had been taken in the art of shipbuilding and instruments like an improved compass and the astrolabe had become part of the equipment of navigation, along with the hourglass and the log. Wealth was the prime motivation. Europe was searching for new markets, especially in the fabled Indies; but the eastern route had been closed by the Turks. If the earth were really round, as Ptolemy had said, there would be little danger of falling off its edge if they were to sail to the west.

After the Vikings, the Portuguese were the next great navigators and explorers. They developed a particularly seaworthy ship called a "caravel." Prince Henry gave royal incentive to exploration by his establishment in 1416 of a school of navigation.

Bartholomew Diaz passed the Cape of Good Hope in 1488, and Vasco da Gama became the first European to journey by sea to India. With four vessels he rounded the southern tip of Africa and sailed across the Indian Ocean to Calcutta in 1497-1499. Out of his voyage grew the Portuguese Empire. Henry the Navigator did not live to see it, but Portugal realized his objective of finding an eastern route to the Orient. The quest for a western passage was left to the Spanish.

At about the same time, the Chinese were making their own voyages of exploration to the west. Cheng Ho made his first voyage in 1405 to the Persian Gulf and the Red Sea. Though there is no evidence that Orientals were seeking a sea route to the New World, a large junk appeared in Europe in 1487.

Christopher Columbus opened up vast new territories for his Spanish backers, although the lands he reached were another ocean away from the Orient he sought. When another of Spain's bold adventurers, Vasco de Nuñez Balboa, stood on a peak on the Isthmus of Panama and gazed at the Pacific, he realized that there were two seas, not one, separated by a continental landmass. There was no middle passage. Then Spain found a navigator willing to attempt a circumnavigation: in 1520 the Portuguese Ferdinand Magellan sailed through the treacherous strait named for him and reached the Pacific.

Giovanni Verrazano furthered the exploration of the coast of the New World and reached what is now Maine in 1524. John Cabot, employed by the English king Henry VII, reached the coasts of Newfoundland in 1497. His son Sebastian went farther north in 1509. The Frenchman Jacques Cartier sailed into the Saint Lawrence River. Sir Hugh Willoughby of England and Willem Barents of Holland both headed due north and lost their lives in the arctic. Henry Hudson, who gave his name to a great river, a strait, and a bay, also gave his life searching for the elusive Northwest Passage. It was not discovered until 1905, when Roald Amundsen reached the Pacific.

All these great voyages inspired new legends, new learning, and new laws.

Cartography. *Every voyage of discovery, even those that failed, contributed to our knowledge of the geography of the world. By 1719 this is the way the world looked to the cartographer.*

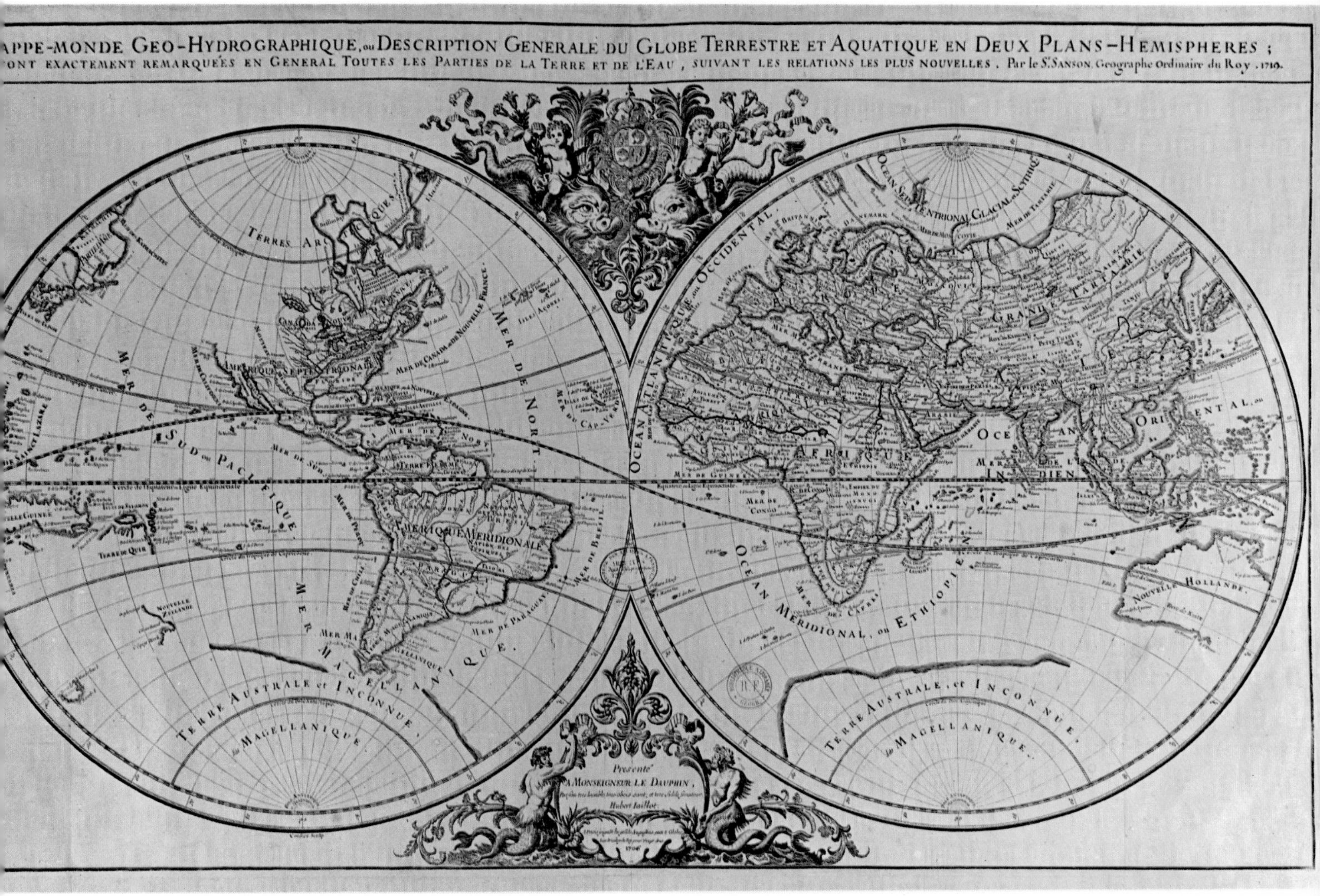

Viking ships *were long open boats called "serpents," usually with a carved figurehead at the prow.*

In the Land Where Wild Grapes Abound

As early as the ninth century, the northern seas were swarming with sturdy oceangoing boats. They were high-stemmed, broad-breasted cargo boats propelled by one large rectangular sail and steered by a long rudder pinned to the starboard quarter. The men who steered them had no compass. They were able to hold course on a latitude by observing the sun and the stars, but they had no way of determining their longitude. These ships were probably the fastest in the world at that time, but they were unable to sail very close to the wind, and often they had difficulty holding course in a crosswind. It was very likely that this circumstance led a young merchant named Bjarni Herjolfsson to be blown off his course as he sailed from Iceland to Greenland in 985 or 986 and to sight unknown lands across the Atlantic.

Fifteen years later, Leif the Lucky, as he is known in the Vinland Sagas, or Leif Ericsson, as he is known to most of us, bought the very ship that had survived this voyage and steered it back to the New World to explore Herjolfsson's chance discovery. What he and his crew found delighted them. Rolling grasslands, meadows full of golden wheat, rivers that teemed with salmon, vast stretches of tall pine, deer and other game in abundance all added up to a new paradise that would hopefully be colonized soon. Best of all, though, they liked the wild grapes that they found growing in profusion everywhere, and so they called the land Vinland—the land of grapes.

These Norsemen had discovered America, but where they landed is still a question. Discoveries recently made, based on calculations of the sailing times, and directions known from sagas, and from an Icelandic map drawn in 1590, support the argument that it was on the coast of Newfoundland.

The Oseberg ***burial ship,*** *dating from 900 A.D., is on display in Viking Ship Hall, Oslo. The 65-foot, shallow-draught oaken vessel has places for 30 oarsmen. Excavated in 1904, the graceful ship contained the remains of a young queen.*

In the Land of Fire

On the Feast of the Eleven Thousand Virgins in 1520, a young Portuguese explorer named Fernao de Magalhaes was at 52° south latitude somewhere off the coast of the New World. The object of his voyage was to find a southern passage to the Indies. But now he was lost.

The weather was uncertain, with scattered squalls and a choppy sea. The bay that his fleet was in looked like all the other bays he and his men had been exploring in vain for many days. Rations were low. Two of the ships were sent to seek a way out. They were hardly gone from sight when a hurricane hit the bay. The flagship of the fleet and the other remaining ship took refuge in the roadstead, but the two that had gone on reconnaissance were already deeply engaged in the strait and could only be presumed lost.

What really happened, however, was that the two ships were driven even further into the bay by the storm. And lucky they were! For suddenly they saw that the bay opened up on the other side as well. They followed the new passage and came to a second bay and from there into yet a third. No other breach had led so far inland as this one, and here the water was salt, and the effects of the tides were strongly felt. They hastened back to inform their commander.

Upon hearing this news, the commander of the little fleet decided to take the risk. Sails were unfurled and anchors weighed, and the four ships gathered way, threading the narrow channel. Desolate, rocky cliffs threatened them from either side. That night they saw the most uncanny sight they had ever seen. All around them glowed countless fires. They did not know that they were passing from one age to another, that they had voyaged back hundreds of thousands of years to a time when men did not yet know how to make a fire and so were compelled to jealously guard a flame day and night lest it should go out forever. All around them were unseen people who still lived as man had lived in the Stone Age. Thus, in this bleak wilderness, hundreds of eerie lights glowed around the four ships. It was a fantastic journey for these men.

Finally they came to a passage within walls of rock as much as 6000 feet high. The sea around them boiled and seethed as if it were angry at this invasion of its domain. Impenetrable mists blanketed visibility. The nautical skill of the commander of the fleet needed a small miracle to negotiate the passage.

But finally it was done, and the fleet found itself in a new ocean. It was pleasantly different from the ocean they knew. It was so calm that Balboa called it the Pacific. The sailors did not know, however, that the hell through which they had just passed would be known by the name of their commander, the Strait of Magellan, and the lands around it, Tierra del Fuego ("land of fire").

The **Santa Clara** *(above and right) is a replica of the* Niña, *the smallest of Christopher Columbus's three ships. It was built by Robert Marx in 1962 and sailed by him in that year from Spain to San Salvador in the Bahamas.*

SANTA

Monsters of the Deep

Over the centuries there have been thousands of reports of the sighting of gigantic and fantastic creatures in the sea and on the shore. Investigations of these reports in recent times have led us to believe that most if not all of the sea monsters described are nothing more than waterlogged trees, masses of seaweed, whales, basking sharks, giant squid, walruses, manatees, dugongs, or seals. Nevertheless, some of the descriptions given have not been all that easy to explain away, and it is not possible to state categorically that strange, still unknown creatures of the sea do not exist. We know that there once lived in the sea animals that we would call monsters if we were to see them today. The fossilized skeletons of some of these denizens of prehistoric seas have been discovered, but most of them are of modest size. The only one able to stir our imagination is the giant Megalodon shark, much bigger than the largest white shark of today. It seems likely that the monsters of the sea are simply creations of the fertile imagination of man.

From the very first time that man ventured out on the high seas, sailors have brought back all sorts of strange, fantastic tales. There were monsters whose backs were a mile wide. Others had arms long enough to pluck a sailor from the crosstrees of a mast. There were even stories of a gigantic serpent that circled the earth at the equator.

That such monsters of the sea undoubtedly do not exist is of little importance. What is important is that the idea of them has given us some of the most exciting, fanciful, and mysterious legends of the sea.

*Of all the **legendary creatures** of the sea, the sea serpent was one of the most terrifying as well as one of the most popular.*

Monsters of the Middle Ages

It took extraordinary courage to be a sailor when the earth was still thought to be flat, for if one ventured too far out to sea he was taking the risk of being swept over waterfalls at the edge. Even if that did not happen, there were always those places in the sea where the water was boiling and where fearsome monsters lay in wait to crush his ship and devour him. The Great Wall Snake was a peril the sailor believed he might have to face at any time. And then, of course, there was always the great serpent that encircled the entire globe at the bottom of the sea with its tail in its mouth.

Naturalists of the Middle Ages did their part to people the sea with imaginary beings. They reasoned that every land animal must have its counterpart in the sea. Thus there must be sea horses, sea dogs, sea cows, sea elephants, and sea lions that looked like their terrestrial cousins, and they were partly right. These early scientists also believed that since the sea was the home of mammals like the whale, the porpoise, the manatee, and the seal, counterparts of human beings might also be found in the sea. Among the species they created and envisioned as living in the sea were the sea bishop and his attendant, the sea monk. The sturgeon might have been the inspiration for these beliefs, because this fish has a body covered with what looks like scaly armor. Perhaps it was a member of the ray family, such as a skate, which has on its underside markings that resemble a human face.

Medieval teratology. *In the Middle Ages some people believed that every person on land had a counterpart in the sea. The sea bishop (left) and his attendant the sea monk (right) are drawings that appeared in* L'Histoire entière des poissons, *published in 1558 by the naturalist Rondelet.*

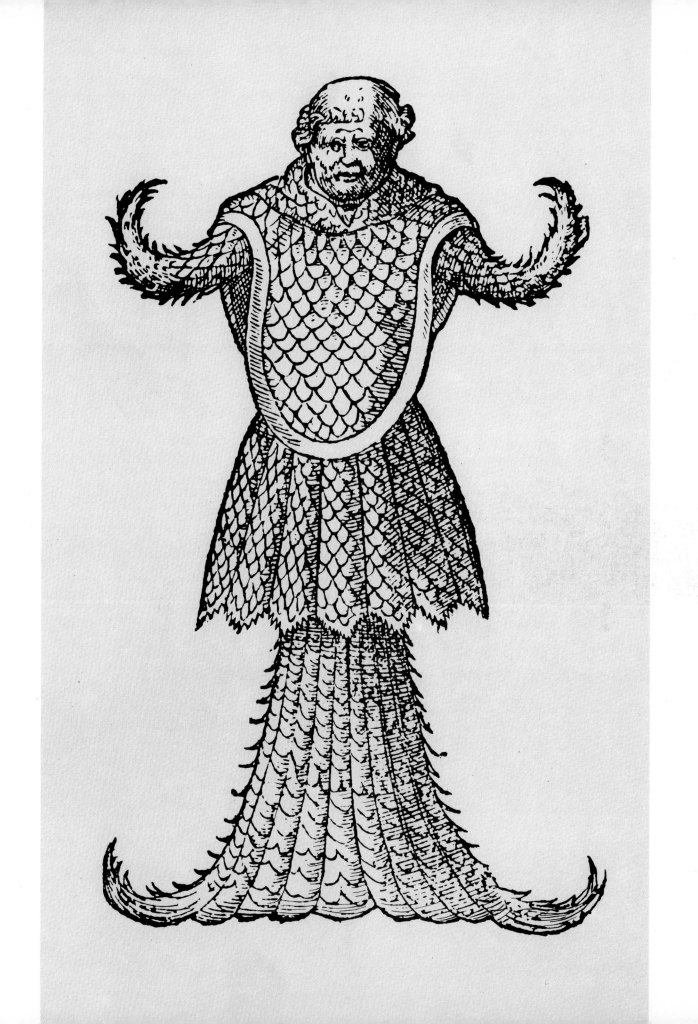

The Kraken

In 1555 a Swede named Olaus Magnus wrote a history of Scandinavia in which he described a peculiar monster called a kraken. It appears that in reality it was a giant squid or octopus or cuttlefish. None of these animals is imaginary. But the imagination of the sixteenth century made the kraken something far more gigantic. Engravings of the time show squids so large that they can snatch a sailor from high up in the rigging of a ship. A famous drawing of a kraken by a Swiss artist who had never even seen the ocean gave the animal a body like a catfish, legs like an alligator, and tentacles that resembled ribbons.

Probably the first recorded mention of something like a kraken was in the *Odyssey*. The picture that Homer gave of the monster Scylla clearly suggests an octopus or a squid enormously magnified.

*The **kraken** (right) was a legendary monster of the Middle Ages that was probably inspired by tales of a giant octopus. A legendary **giant squid** (below) attacks a diver.*

The Loch Ness Monster

No monster past or present has had more newsprint devoted to it than Nessie of Scotland's Loch Ness. Monster sighting at Loch Ness is a perennial sport that reaches its highest pitch during the holiday season each August. It is then, during what they call "the silly season," that the editors trot out the Loch Ness monster story.

Those who believe in Nessie explain its presence in landlocked Loch Ness quite convincingly. During the last ice age when the weight of the ice had pushed down on the earth, Loch Ness was connected with the sea. As the ice melted, the sea level rose, but without its heavy cover Scotland rose too, like a cork in water. Loch Ness was cut off from the sea, and many marine animals were trapped in it, among them Nessie's ancestors. The long, narrow loch, averaging 430 feet and plunging at one point to 754 feet, could support a large animal in its deep water.

Sea monsters. *As it made its way through an inlet connecting the lake with the ocean at the end of the last glacial age, the Loch Ness monster might have resembled this horrendous creature in a painting by the Florentine Piero di Cosimo (1462–1521).*

Nessie's notoriety goes back a long way. In 565 A.D. it had killed one boatman and was chasing another when the abbot of Iona, Saint Columbia, came along. The abbot raised his hand and moved it to make the sign of the cross. Then he commanded the monster to cease and desist, which the terrified beast immediately did. The abbot's spell was a lasting one, for there have been no reports of molestation from the animal again.

A lot of people say they have seen Nessie, who likes to bask in Scotland's thin sunlight and might be mistaken for a floating log. She is estimated to be a fast swimmer, reaching 10 to 15 knots. One Englishman, Tim Dinsdale, has made a literary career out of Nessie, as a result of 13 years of monster-sighting. A retired gentleman named Frank Searle at last report has clocked more than 15,000 hours of it. In recent years Nessie has been sought by divers and even by men riding in a yellow submarine and by people peering out of a submerged capsule. Probes have been made electronically with echo sounders. But Nessie is resistant to discovery; it lives in water where visibility is measured in inches—a perfect habitat for bashful beasts.

An ambitious Japanese impresario, Yoshio Kou, led an investigating party to Loch Ness in 1973. Kou believed there were a lot of monsters, and he planned to tranquilize and capture a few. The scheme came to nothing when the chief constable heard of it, and he made Kou promise not to attempt it.

Skeptics say there isn't any Nessie. To settle the matter, an important British peer suggested, not entirely in jest, that the loch be drained. That would be bad for tourism and make a lot of people unhappy.

Scotland is not the only country that can boast of such a creature. Ireland has a great variety of them. Iceland has its *skirmsl,* a 46-foot lake monster. Lake Victoria in Africa has its gigantic *lau.* Canada's Lake Okanagan allegedly harbors a monster called *ogopogo* by the Indians, and to this day they refuse to cross the lake in certain places.

"The ocean . . . offered us an incessant and infinite display of its most marvelous treasures."
Jules Verne,
Twenty Thousand Leagues under the Sea

A Letter from Dieppe

"I wrote the other day saying that my mind is occupied by improbable notions. But in fact, they are not improbable at all. Whatever one man is capable of imagining, other men will prove themselves capable of realizing." The man who wrote these words in a letter to his father from Dieppe in 1868 was writing the greatest imaginative work on exploration of the depths of the sea. His name was Jules Verne, and the book was *Twenty-Thousand Leagues under the Sea.*

Verne was not a scientist, but he was a diligent student of science. He took notes on anything that had to do with science, explorations, discoveries, inventions, travels, and adventures. It was this that helped to make the *Twenty Thousand Leagues under the Sea* plausible. Here is a world of wonderful underwater plants, coral forests, fantastic creatures of all kinds, monsters, volcanoes, the polar regions, cemeteries, the ruins of Atlantis, coal mines, treasures—all seen on the voyage of a submersible.

The commander of the underwater vessel, the *Nautilus,* which is both a scientific laboratory and a refuge from the rest of mankind, is Captain Nemo, who loves the sea not only for itself, but because it symbolizes all he values most: "The sea is everything. Its breath is pure and healthy. Here man is never lonely, for on all sides he feels life astir. The sea does not belong to despots. Upon its surface men can still make unjust laws, fight, tear one another to pieces, wage wars of terrestrial horror. But at thirty feet below their reign ceases, their influence is quenched, and their power disappears. Ah, sir, live—live in the bosom of the waters. There alone I recognize no master. There I am free!"

Through all the seas the *Nautilus* travels with its enigmatic captain and its crew, Professor Arronax, the faithful Conseil, and Ned Land, the harpooner, giving a glimpse of wonder after wonder, never entirely real or imaginary. As its narrator says, "the ocean, during this cruise, offered us an incessant and infinite display of its most marvelous treasures. There was a continuous change of decor and scenery, as if staged to please our vision, and we were called upon not only to contemplate the works of the Creator in this vast expanse of liquid world, but also to delve into the most redoubtable mysteries of the sea."

For Verne, the sea was a living creature. As Captain Nemo says: "Look at this ocean! Is it not endowed with a life of its own? Does it not have its moods of anger and tenderness? It has a pulse and arteries. It has respiratory organs and a circulation just as truly as we have a circulation of the blood. To endow it with life, the Creator of all things had only to bring together, in the sea's depths, heat, salt, and animalcules. Heat creates different densities, which start currents and countercurrents in motion. Evaporation, which is nonexistent in the extreme polar regions but which goes on at a great rate in the equatorial zones, means that there is a continuous exchange between tropical and polar waters. I have detected these downward and upward moving currents which are the true respiration of the ocean. I have seen its infinite, exuberant life permeating every molecule of the liquid element."

"It was an oyster of extraordinary size, a gigantic clam that could have been made into a holy-water basin..."
Jules Verne,
Twenty Thousand Leagues under the Sea

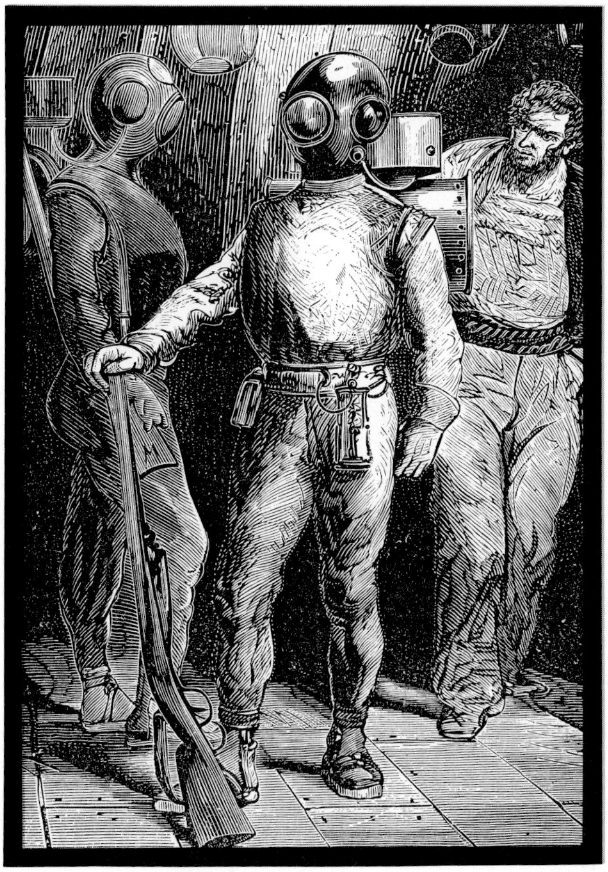

"With my Ruhmkorff lamp hanging from my belt and gun in hand, I was ready to go."
Jules Verne,
Twenty Thousand Leagues under the Sea

Fantasies like these, with which the book is strewn, are so rich in conjectures about the nature of the sea that even today scientists are putting to the test some of the questions they raised. Verne, like many great poets, led the way for science to follow.

1
7
5
6
2
3
8
4

With Darwin on the *Beagle*

"During our stay, I observed the habits of some marine animals. A large Aplysia is very common. This sea slug is about five inches long; and is of a dirty yellowish color, veined with purple. On each side of the lower surface, or foot, there is a broad membrane, which appears sometimes to act as a ventilator, in causing a current of water to flow over the dorsal branchiae or lungs. It feeds on the delicate seaweeds which grow among the stones in muddy and shallow water; and I found in its stomach several small pebbles, as in the gizzard of a bird. This slug, when disturbed, emits a very fine purplish red fluid, which stains the water for the space of a foot around. Besides this means of defense, an acrid secretion, which is spread over its body, causes a sharp, stinging sensation, similar to that produced by the Physalia, or Portuguese man-of-war."

This is a characteristic observation that Charles Darwin made and recorded during his five-year voyage on the *Beagle*. It is this kind of attention lavished on so seemingly inconsequential an animal as a sea slug that was to lead him—and us—to an understanding that we are not isolated from nature. We are, indeed, a part of nature; we are all descended from more primitive creatures and, ultimately, from the common origin of all life on earth. Some years later Charles Darwin was to formally present this view in his *Origin of Species*. His theory was not without its detractors.

Recorded observations, *painstakingly made from nature, led to classification of all plants and animals. A plate (left) by a nineteenth-century naturalist illustrates dates and dolphins.*

The close attention ***Charles Darwin*** *paid to such seemingly inconsequential animals as the sea hare (below) enabled him to develop his theory of the common origin of all life on earth.*

No one knows why these ***monolithic heads*** *were erected, facing the ocean on Easter Island in the South Pacific, or who made them.*

The Undeciphered Tablets

Ever since it was discovered by the Dutch in 1722, Easter Island, that tiny, remote speck in the South Pacific, has been surrounded by a halo of mystery. Its colossal stone statues have been subject to speculation as well as scientific inquiry. There are more than 600 of these monoliths, strikingly similar eyeless faces of men or gods. Most are between 12 and 25 feet in height, but some measure as much as 40 feet and weigh up to 70 tons. Recent investigations, most notably that by Thor Heyerdahl, have uncovered many hitherto unknown statues. A kneeling colossus was unlike anything previously seen. Their presence on this solitary and bare islet is as much of a riddle today as it was when the first European to contemplate them wrote in his log: "These stone figures filled us with amazement, for we could not understand how people without solid spars and without ropes were able to raise them."

Perhaps even more intriguing than the statues are the many wooden tablets that have been found on the island and that bear hieroglyphic writing that still has not been deciphered.

Where did the people come from who built ***the statues of Easter Island?*** *One theory maintains that they migrated to the west from present-day Peru.*

冨嶽三十六景
神奈川沖
浪裏

Kon Tiki

On April 28, 1947, all flags were hoisted on a fragile craft ready to make a voyage of discovery—a raft. Its name was *Kon Tiki* and its commander was an intrepid Norwegian named Thor Heyerdahl.

Heyerdahl was attempting to provide evidance that the ancestors of present-day Polynesians could have emigrated not from the west, as it is still believed but from the east, possibly from Peru. Legend said that it was so. The sun king, after whom the raft was named, was said to have led his fair-skinned people westward to Polynesia more than 1500 years ago on balsa rafts just like Thor Heyerdahl's.

Heyerdahl has told the exciting tale of the voyage in his book *Kon Tiki*—of catching sharks barehanded, of quenching thirst by eating raw fish, of the frequent visits of curious whales that gave the men some uneasy moments. The men of the *Kon Tiki* were the first to see a living snake mackerel. The unusual fish jumped on board one night and slithered into a sleeping bag with one of the crew. For a fish to come aboard was not unusual. The cook's first duty every morning was to collect all the flying fish that had landed on the deck during the night.

When Heyerdahl and his five companions reached the Tuamotu Islands, sighting the atoll of Puka Puka on July 30, the group had given the argument of a westward migration evidence of its plausibility. But the scientific world was not convinced, and the mystery of the origin of the peoples of the South Pacific remains.

The Breaking Wave off Kanagawa, *by the Japanese artist Hokusai (1760-1849), shows three fragile boats about to be smashed by an enormous wave. Mount Fuji appears in the background.*

Chapter V. The Birth of Venus

When man got over his great fear of the ocean and the anger this fear provoked, he was able to become sensitive to beauty inherent to the sea, and to the gorgeous shapes and colors of marine creatures. Venus could only be born from the sea, but the sea had to produce also a beautiful being more popular, more accessible than a goddess.

> **"Her skin was as clear and opalescent as a rose petal. Her eyes were as blue as the deepest sea. But like all the others, she had no feet. Her body ended in a fishtail."**
> **H. C. Andersen**
> **"The Little Mermaid"**

In the north, that is to say, in the land of the Norsemen, the Vikings, who were the first to overcome their terror of the sea, invented the most charming and enduring creature ever to grace the waters—the mermaid. She was so immediately popular that soon the whole world was talking of her. By the Middle Ages the churches and even the cathedrals displayed her image in wood and stone. Of course, the fascination that the church took in her was less than one of sheer rapture. She was also dangerously alluring to the infatuated sailors.

In the Mediterranean, the idea of the mermaid was so seductive that the siren shed her bird form and grew a fish tail. Presently, she acquired those appurtenances that have remained so necessary right to the present day —her comb and her mirror.

The stories that grew up about her could have filled the old library at Alexandria. In many of them the mermaid longs to possess a soul which, in her present existence, is denied her. Without a soul she has no hope of salvation or an afterlife, and this prospect causes her much distress.

The accounts of voyages are often filled with their tales of mermaids. Christopher Columbus, however, allots them only a brief space. The mermaids he saw failed to come up to his expectations.

The mermaid's history shows her to be a beautiful, enticing, occasionally prophetic sprite with a mellifluous voice. Inspired by the sea, the mermaid in turn has inspired countless poets, sculptors, and painters. Even composers have written music to her.

Eventually, as man began to take his reasonableness more and more seriously, belief in the mermaid began to wane, but interest in her was unflagging. Men sought to explain her origin and persistence in rational terms. It was said she was, in fact, a dugong or a manatee or a seal basking on a rock. These hypotheses had one good result—they encouraged men to take a closer look at these animals and to observe the characteristics that supported the theory. But it must be remembered by anyone who would attribute the mermaid to mistaken identity—a sailor has a keen eye. He knows a dugong or a manatee or a reclining seal when he sees one. No, the mermaid filled a very deep emotional need for tenderness and beauty for those who had to endure the terrible rigors of the ocean. It's only a shame she isn't real.

***Botticelli's* Venus,** *the goddess of love, emerges from a giant scallop at her birth. Shells provided the geometric perfection needed for such elegance.*

XAP

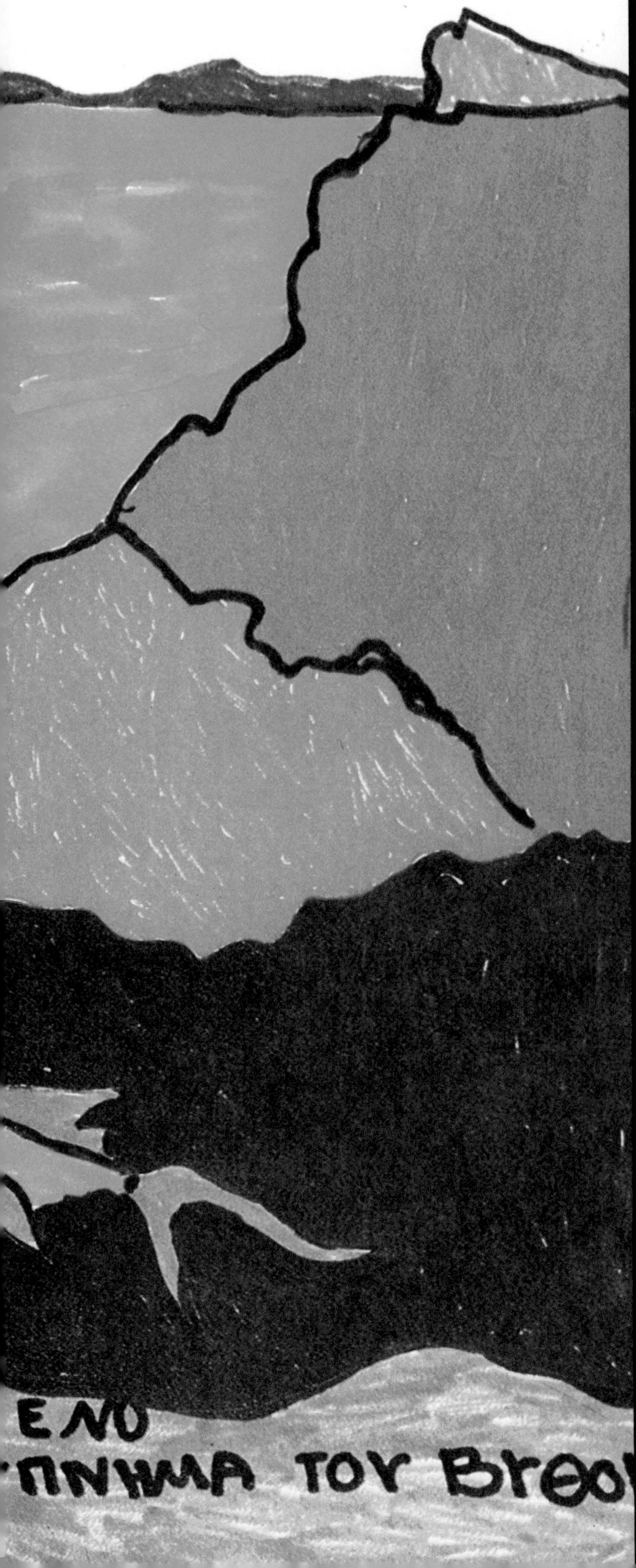

How the Fish Got into the Sea

Once there were fish who lived on land and were great hunters of other animals. The fish would still be living on land if a terrible thing had not happened to them one day. A group of them had just returned from a hunting expedition and were cooking their food when a great black cloud, unlike any they had ever seen before, blotted out the sun, and a chilling rain began to fall. The fire was all but put out by the rain and, as you know, that is a very serious thing where people have no matches.

Just then the youngest of the tribe said, "Let my father, the cod, light the fire. He is very skilled in magic—more than most fishes." The big cod stripped some bark off a tree and placed it on the smouldering ashes. Then he knelt down beside it and blew on it. Soon the spark had grown into a flame, and a happy crackling was heard. "Soon we shall be warm again," the delighted fishes said to one another. "Truly the cod is great!" Suddenly, with a shriek, a blast of wind swept down from the hills and blew the fire toward them. They sprang back and, quite forgetting where they were, tumbled in a heap over the bank into the water below. They were freezing in the dark water where the sun never shone, but the wind blew the fire right in after them. And the fishes gathered around it just the way they had on land, and the fire burned under the sea as brightly as ever. In fact, they never had to light it again, and it burns to this very day. That is why if you dive below the cold surface of the sea you will find it pleasantly warm, and you will be sorry that you cannot stay longer.

A child's sea, *as painted by a Greek child. Although naive, it is no less true or meaningful than any photograph or description in words might be.*

Why the Fish Do Not Speak

The national epic of Finland, the *Kalevala,* a legend that is well known to all Finns, explains why the fish do not speak.

There was a time, the legend goes, when no one knew how to speak. The animals had no cries. The birds had no songs. The waters flowed and the winds blew, but they made no sound. Even man made no sound.

One day Vainamoinen, the Master of Song, sat down with his harp, and when he plucked its strings, every creature on earth or in the sky pricked up its ears, and so did the wind and the waters and the trees. And Vainamoinen commanded them to take for their own the language that suited each best.

The wind chose the loud roar and rattle of Vainamoinen's big boots as he mounted to his seat. But the thunder got first choice, so the language of the thunder is much louder than the wind's, but on the other hand it never talks for as long a time as the wind does. The river decided that the rushing swish of Vainamoinen's cloak made a delicious sound. The trees thought the rustle of Vainamoinen's sleeves was best for those who had leaves for lips. The birds were fussier than the others, and they found no speech pleasing until Vainamoinen played a little melody on his harp. All the creatures of field and wood, desert and jungle gathered around the Master of Song. As they listened, each discovered a way of whistling or humming, barking or bellowing, that seemed a proper language. As for man, he learned all the different sounds that Vainamoinen's harp made and that his garments made as he moved this way or that. And he learned to sing better than the birds themselves.

But while everything on the earth and in the sky had been listening to the great Master of Song and choosing a specific language for itself, the fish had been quite helpless. They knew that something of great importance was going on, but they had no idea what it was. The fish could see all the creatures of earth and sky opening their mouths and shutting them but, being underwater, they couldn't hear a sound. Nevertheless, they made up their minds that they should behave just like the others. That is why you can see the fish opening and closing their mouths and not making a single sound.

Paul Klee *(1879-1940) was one of the most original and inventive painters of the twentieth century. He saw in primitive art and children's drawings what he was always seeking for himself—elementary forms uncompromised by convention. His* Fish Magic *is shown at left and in detail below.*

Why Fish Have Gills

There are many legends that tell why fish have gills, and a favorite in the islands of the South Pacific concerns a grouper that fell in love with a beautiful girl.

The fish saw her one day as she was weaving on the shore, and he fell in love with her at first sight. He set about wooing her in the only way he knew—by swimming about as fast as his sluggish body would permit and splashing water at the girl with his tail and fins. Twice he asked her to marry him, and twice the girl refused, and he swam out to the deep waters of the reef and lay there mourning for his unrequited love.

Then he decided that he could not give up so easily, so he swam back into the lagoon right up to the shore where the girl was weaving. He stretched out a fin, snatched the girl, and flipped her into the water. Before she knew what was happening, he swallowed her. Of course, he was careful not to hurt the girl for he loved her very much. When the girl realized where she was, she demanded to be let out, but the grouper refused. "I love you," he cried, "and I cannot let you go."

Then the girl had an idea. She still had with her the sharp shells that she used to pattern her tapa cloth. With them, she cut two slits in the fish's body, one on each side, and then she slipped out and swam to shore.

As for the grouper, once he got used to the water rushing in and out of his throat through the cuts, he rather liked the feeling, but he vowed never again to fall in love with a human girl. And that is why all fish have gills in their throats today.

*According to legend, the **grouper** (above) has gill openings on the side of its head because he swallowed a beautiful girl he was in love with who then was forced to cut her way to freedom through the side of the fish. He adjusted to the flow of water through these openings and retains them to this day, but he vowed never to fall in love again.*

*The **jack** (right) has obviously not taken heed, as did the grouper, to the pitfalls of falling in love with and swallowing pretty girls.*

Why Jellyfish Have No Shell

From Japan comes a tale that explains why jellyfish have no shell. A long time ago it happened that the Sea King's wife fell ill, and the doctor told him that the only thing that would cure her was the liver of a monkey. One of the few denizens of the sea that was able to walk on land was the jellyfish, and to him was given the task of journeying to Monkey Island and enticing one of its inhabitants to return with him. Obediently, the jellyfish set out. When he landed, with the help of a convenient wave, on the shore of the island, he spied a monkey in a tree and fell into a conversation with him. Soon he was telling the monkey of the splendors of the palace of the King of the Sea, of its trees of white, pink, and red coral, and of the fruits that hung from their branches like great jewels. The monkey was so entranced that he agreed to go with the jellyfish to see these wonders for himself, and he climbed onto his back and they set out over the sea.

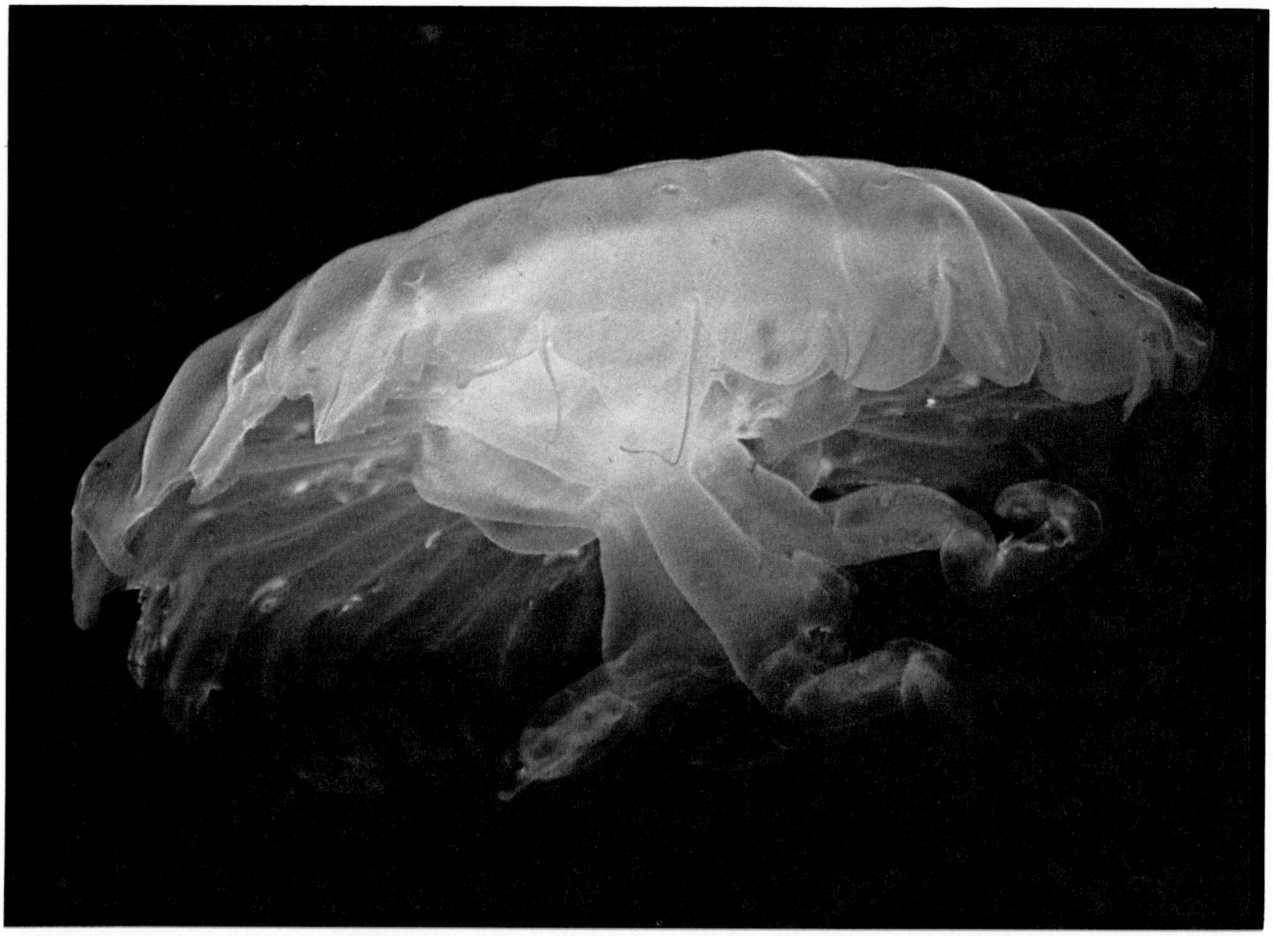

Presently, however, the jellyfish, who had little knowledge of anatomy, asked the monkey if he had brought his liver with him. The monkey was perplexed at such an unusual question and demanded to know why the jellyfish had asked. The jellyfish, suddenly feeling sorry for the monkey, told him everything. The poor monkey was horrified at what he heard and angry at the trick that had been played on him. He trembled with fear at the thought of what was in store for him. But he was also a very clever monkey, and he thought that the wisest plan would be not to let his fear show but to calm himself and try to think of some way to avert such an awful fate. Then he had an idea. He told the jellyfish that he had not brought his liver with him but had left it hanging on the tree where he had been found. So they turned around and swam back to the island. Just as soon as they landed on the beach, the monkey scampered up into the branches of the tree and jeered at the jellyfish and admitted that he had deceived him. There was nothing the jellyfish could do now except repent of his stupidity at letting the monkey dupe him and return to the King of the Sea and confess his failure. Sadly and slowly he began to swim back. The last thing he heard as he glided away was the monkey laughing at him.

The king's wrath was great, and he at once gave orders that the jellyfish was to be severely punished. His bones and his shell were taken from him and he was beaten to a pulp with sticks. Then his limp and battered body was carried out beyond the palace gates and he was thrown into the water. Ever since that day, the descendants of the poor jellyfish have been soft and boneless.

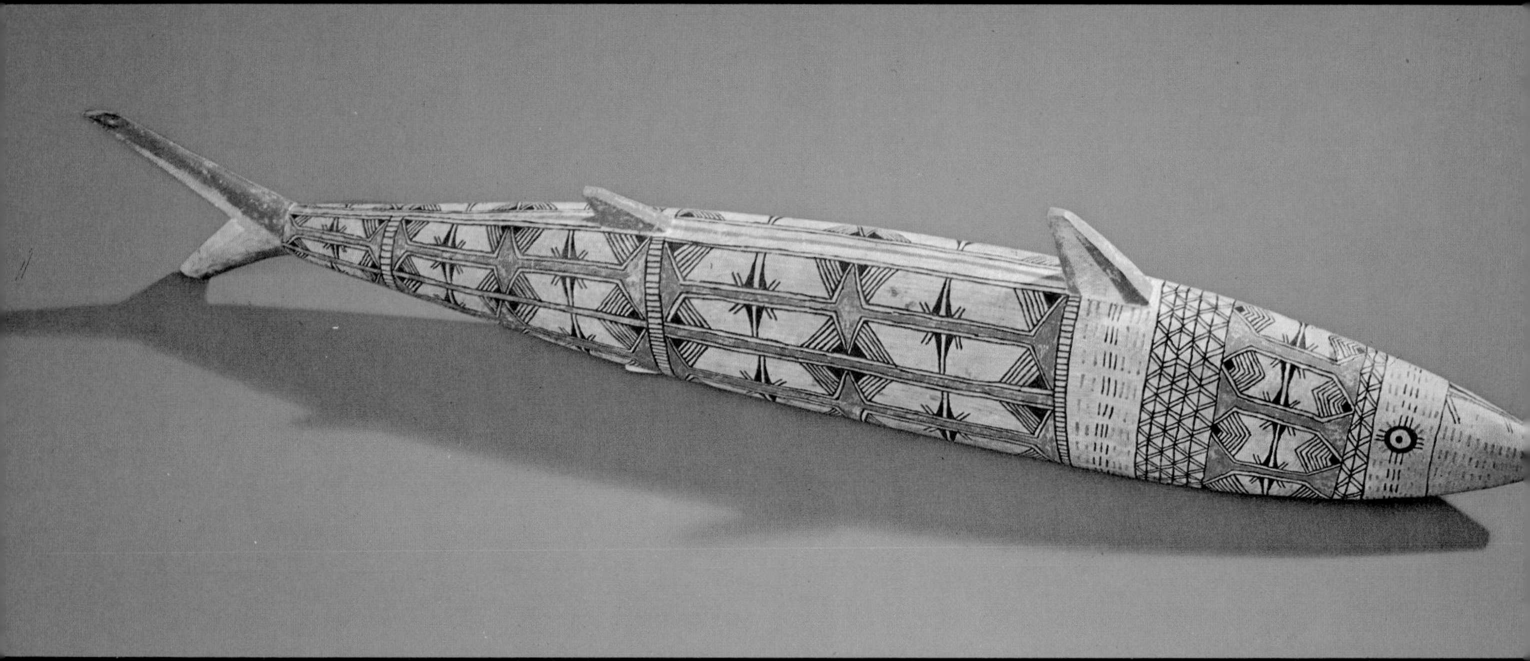

Why the People of Raiatea Have No Fear of Sharks

One day it happened that the lazy octopus who guarded the lagoon of Raiatea fell asleep, and sharks got in and ate ten of the islanders. The chief of the island, whose name was Covered-with-Scars, was very troubled, and he asked his son for advice. The boy was wise and cunning. He reasoned that the invasion of the sharks had come about because the giant octopus that guarded the lagoon had fallen asleep. He knew it would do no good to wake him up, for he would soon fall asleep again. The boy thought about it for some time, and then he came up with a plan. He swam far out into the lagoon and floated around until he saw a big shark that had come in past the sleeping octopus. The boy told him that he had a message for the King of the Sharks. The shark thought that sounded unlikely, but he wanted to be sure. So he told the boy to get on his back and he would take him to the king. The boy climbed on the shark's back and held tightly to its dorsal fin. The shark swam swift as an arrow out into the open sea and straight to the king, who lay on the bottom, all covered with barnacles, sand, and seaweed.

The boy told him that the Turtle of Tahaa, who guarded the entrance to the lagoon of that island, had challenged him to battle. The King of the Sharks was delighted to accept the challenge, and he swam directly to Tahaa and swallowed the turtle in one gulp. Then the boy told the Shark King that the giant eel that guarded the lagoon of Bora Bora had also challenged him. The king went straight to that island where he easily devoured the eel.

Then the boy told the King of the Sharks that the giant octopus that guarded the lagoon of Raiatea had challenged him. The shark swam directly to Raiatea and, in a voice that could be heard for miles, shouted at the octopus until he woke him up.

Immediately the octopus wrapped four of his legs around a rock buried in the sea bottom to give himself a firm anchor. Then he stretched his other four arms across the passage in the reef and waited for the attack.

It came at once. Like an angry black shadow flitting through the depths, the Shark King swam at the octopus. Instantly the octopus wrapped four strong arms around the body of the shark and began to squeeze. Then he brought up his other four arms and wrapped them around the shark. Soon the great fish was gasping and groaning and thrashing about, trying to escape the grip of the octopus, but to no avail. The eight arms of the octopus only squeezed tighter and tighter. The battle went on for seven hours, and at one point it became so fierce that it caused the whole ocean floor to tilt slightly. At last the shark could endure no more, and he pleaded with the octopus to let him go. The boy made the Shark King promise that his tribe would never again bother any man, woman, or child of Raiatea, and the shark was only too glad to agree. So the boy ordered the octopus to let the shark go free.

The King of the Sharks carried the boy on its back all around the lagoon of Raiatea to spread the word of the promise. And the lazy octopus, finally relieved of his duties as guardian of the lagoon, yawned widely and went back to sleep.

Shark carvings from the South Pacific. *The shark has always been one of the most feared inhabitants of the sea. Perhaps it has been to ease that fear somewhat that it has been one of the most popular creatures of the sea depicted in art, particularly the art of the South Pacific from which these magnificent painted carvings come.*

Hawaiian shark legend. *At the entrance to Pearl Harbor in the Hawaiian Islands there once lived a famous shark goddess named Kaahu-pahau and her brother, "Smiting Tail," who guarded the waters against man-eating sharks. They had once been humans themselves, but they had been transformed by the gods, and so they were still friendly to the people of the islands, who fed them and scraped their backs clean of barnacles.*

The Boy Who Was Caught by a Clam

In the Fiji Islands you can hear a tale of the boy who was caught by a clam.

One day the boy, who had charge of all the chickens on one of the islands, swam out to the reef and discovered juicy worms and lots of other things that chickens love to eat, and he called to them to come out and join him. They did so and fell to eating their fill.

Soon the tide began to rise, and the boy was just about to warn his chickens when, all of a sudden, he stepped between the open shells of a giant clam. The clam, annoyed at the intrusion, immediately closed its shells, trapping the boy's foot inside.

The boy remembered that people always said that he was gifted with a clever brain and a quick tongue, and he thought he might outwit the clam. So he told his captor that he was grateful to him for having taught him a lesson. Henceforth he would be more careful where he stepped. But this did not appease the clam, who was angry at the boy for bringing his chickens to the reef to feed.

"Those chickens aren't sea creatures," he said. "They belong to the dry land. And so do you. This is my kingdom out here." And he swore that he would hold the boy fast until he drowned.

The tide rose steadily higher, and the boy commanded the chickens to fly back to the island. Then he had a brilliant idea, and he said to the clam: "I thought you were a clam, but I have been mistaken, for I never knew a clam to have a pearl inside him." The clam was flabbergasted. "You're not a clam at all," the boy went on, "you're an oyster!"

The clam was indeed angry, as well as a bit confused, for he had always thought that he was a clam. "Me, an oyster? By no means! I have no pearl inside me! Look!" And he parted his two shells to prove to the boy that he was wrong. And when he did, the boy pulled his foot right out and swam for all he was worth for the shore. His flock of chickens called to him, urging him to hurry to escape the rising tide.

Ever since that day, everyone in Fiji knows when the tide is beginning to rise—for all the fowl begin to crow.

Giant clam, Tridacna *(below), possesses a large mantle (closeup, opposite page) that completely hides the shell from a diver's view.*

Why the Sea Is Salt

Way down at the bottom of the ocean is the source of all the salt in the sea. There once had been a king of Denmark who was very happy because all his people loved him. There was only one thing that bothered him. He had among his treasures two enormous grinding stones. The stones could grind jewels and anything the king wished for. But they were too heavy for anyone to turn.

Once, when the king was in Sweden, he saw two women who had been captured in the land of the giants. The king of Sweden gave them to him as a gift. He took them home and put them to work turning the grinding stones. He commanded them to grind gold and silver and peace and joy, and the giants obeyed. But one day they got tired and asked the king if they could rest. The king insisted that they grind on, and so the women decided to play a trick on him. They stopped grinding gold and silver and peace and joy and began grinding soldiers for the king's enemies. When his enemies were strong enough, they attacked the kingdom and carried off the magic grinding stones and the two giants. The king's enemy had a great need of salt in his country, and while they were still aboard the ship carrying them to that land, he commanded the giants to grind salt. And they did. They ground and they ground, and the salt filled up the ship. Finally there was so much salt that the ship sank to the bottom of the sea. All of its company was drowned except the two giants, and they went right on grinding salt. Even today no one has told them to stop, and that is the reason the sea is full of salt.

The Celts tell of a baby who was abandoned in the sea and of ***gulls*** *that brought him back to land and made a bed for him of their feathers. An angel gave him a bell, and every day a doe gave him her milk.*

The Origin of the Coconut

Did you ever wonder where the coconut came from? If you ask those who live in the islands of the South Pacific, you might hear a story something like this—the lovely tale of the Turtle of Tamarua.

The legend tells of an 11-year-old princess and the Prince of the Turtles, who saw her bathing one day and fell in love with her. The Turtle Prince wanted to marry her, but the girl insisted that she was still too young. Even when the Prince of the Turtles transformed himself into a handsome young man, the girl refused, for she could not leave her dear old father. The Prince wept large turtle tears to hear this, but he was not angry. In fact, he made the girl a present. He told her that when she got home, it would begin to rain, and it would continue until the waters were up to the doorstep. In the morning, the girl would find a turtle outside the house. She was to take her father's ax and chop off its head. Then she was to bury the head and

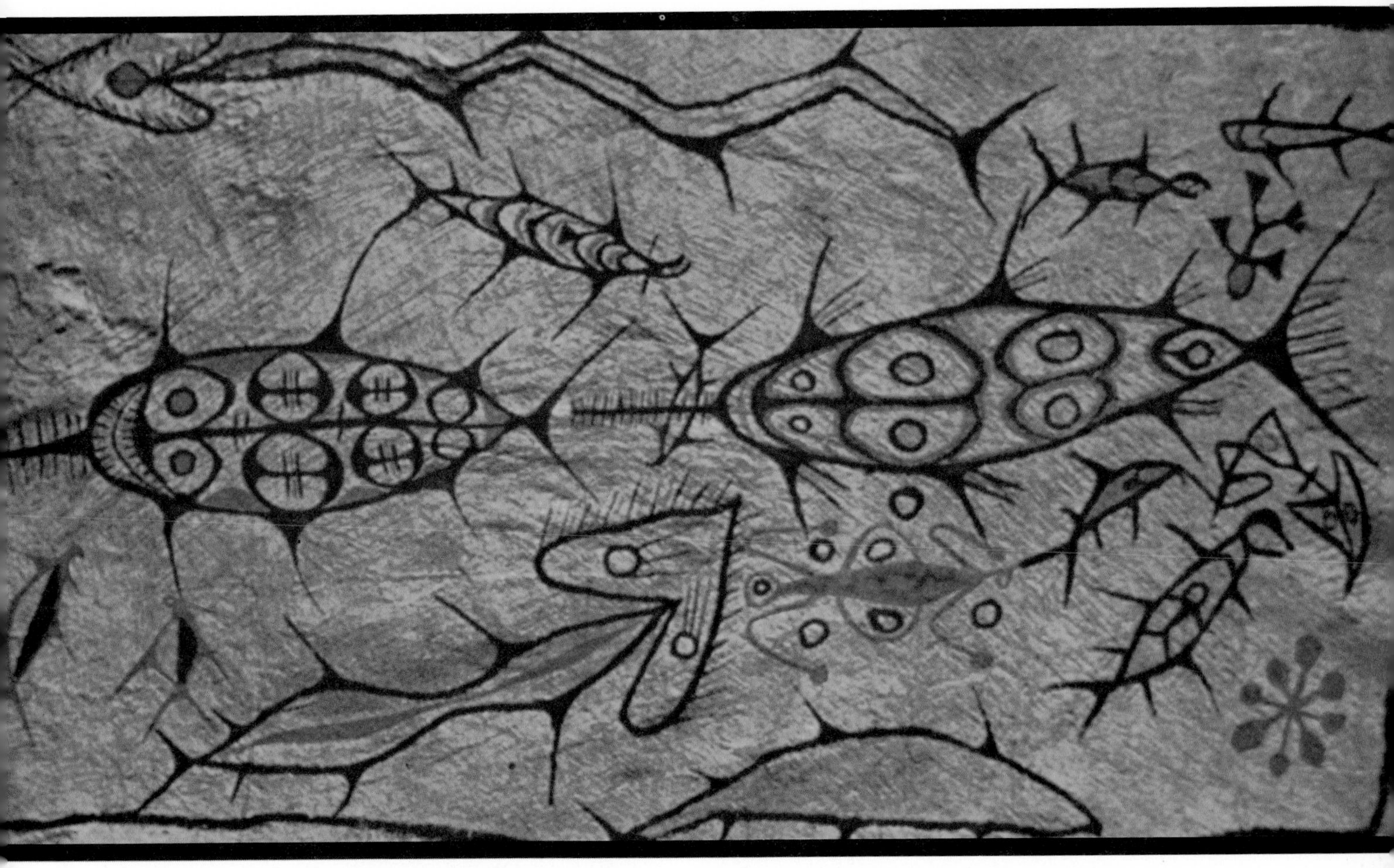

If it were not for a ***faithful turtle,*** *the island of San Cristobal in the Solomons would long ago have sunk into the sea. Legend says that when an earthquake occurs the turtle tightly clasps the rock at Haununu that holds the island up. The painting of a variety of creatures of the sea on barkcloth (above) also comes from the Solomon Islands.*

the body together on the hillside above her father's house.

"Will the turtle be you?" asked the girl.

"No," answered the Prince of the Turtles, a bit haltingly. "It will only be one of my messengers." When the girl got home, it began to rain, just as he had promised. And in the morning there was the turtle, and the girl killed it as she had been instructed and buried it on the hillside.

A few days later a short green shoot sprouted from the grave. Day by day it grew taller and taller, and soon the people of Tamarua could see that it was to be a plant unlike any they had ever seen before.

When the shoot had grown into a tall tree, it flowered and bore fruit, and the girl knew what her gift from the Turtle Prince was to be. It was a coconut—a truly wonderful gift, not only for her, but for everyone in the South Pacific. Ever since that time, the people have eaten its meat and drunk its milk and used its tough leaves for weaving mats, baskets, fans, and thatched roofs.

If, perchance, the reader does not believe in the truth of this tale, he has only to take a look at a coconut to see that its shell is so hard as to be almost unbreakable—just like the shell of a turtle. And the milk inside the coconut is clear and limpid—just like the Turtle Prince's tears.

The Gratitude of the Fish

On a small island in Japan, where nearly everyone is a fisherman, there once lived a man who treated the sea with kindness and was rewarded for it.

Hikoichi was still a boy when he was taken on as a cook by the captain of one of the fishing boats. He got the job only because no one else wanted it. No one wanted to work in a hot, steaming galley, cooking rice and chopping vegetables, while the boat pitched and rolled in the middle of the sea. Besides, everyone considered Hikoichi to be too stupid to ever become a fisherman, but not too stupid to learn to cook.

The fishermen on this particular boat ate well, and there were always a lot of leftovers.

Hikoichi threw them into the sea, but when he did, he called to the fish to come and have a good dinner. They always came and devoured every morsel he threw overboard.

The fishermen laughed when they heard him. "Listen to the fool talking to the fish," they said. And one of them said mockingly, "Perhaps one day they will answer you and tell you how grateful they are!" Hikoichi never paid any attention to the others as they made fun of him but just went on feeding the fish every day.

The ship was soon far out at sea and it was late at night. The men were tired from a long day's work and not long after dinner they were sound asleep. Hikoichi, however, was still in the galley washing the rice for the next day's breakfast. All of a sudden he noticed that the ship felt very peculiar. It was no longer rolling. In fact, it seemed to be standing perfectly still. He went up on deck and peered over the side. He could not believe what he saw. The ocean was gone! There was not a single drop of water anywhere, and the ship lay on an endless beach of shining golden sand. "Why, this is beautiful," he said to himself. Then he decided that he must have some of the lovely sand to remember this night by, so he filled up a bucket with the glistening sand and took it back to his cabin. In the morning Hikoichi was the only one who was surprised to find that the ship was on the ocean, and he told the others what he had seen the night before. Needless to say, they all thought he had simply drunk too much sake, until Hikoichi brought them the bucket of sand as proof that he had not imagined it all.

When the others saw the bucket of sand, their mouths fell open and their eyes stared wide, for it was not a bucket of sand at all, but a bucket full of gold. The next time the ship put to sea, Hikoichi was no longer aboard, for he was wealthy now and lived in a beautiful new house. He bought himself a small boat, and every day he went out to sea and fed the fish. He used his treasure from the sea wisely and well, and he lived a long and happy life on the island where no one ever called him stupid again.

Whaling *has long been an important industry for the Japanese. These paintings are from a scroll that depicts the whaling process through all its stages.*

The Mermaid of the Magdalens

There's a Canadian tale of a young girl taken from her home by a lobster and turned into a mermaid. This happened as a result of a slaughter of sardines by fishermen.

The sardines asked the big fish to help them. In answer to their appeal, a meeting of all the fish in the sea was called. The big fish took an oath to help their small cousins in their struggle with man and to punish anyone who ate or fished the sardine family.

One day a large ship loaded with packed fish was wrecked on the sunken rocks of the Magdalen Islands, off the northeast coast of Canada. The ship was broken up by the heavy surf and the cargo strewn along the shore. That evening, after the sea had calmed, a young girl walked along the shore to view the wreckage of the broken ship. On the beach she saw one of the boxes of sardines, and she resolved to eat them. She tried to smash the box against a rock, but it

*In **French legend,** the best known mermaid is Mélusine, who married a mortal, Count Raymond, and made him promise not to visit her on one certain day each week. Count Raymond did not know that Mélusine was a mermaid, for she had always disguised herself from him. But one day each week she assumed her true form. When he broke his promise and discovered her secret, she fled. Thereafter, her appearance or cries were said to foretell the death of the lord of the castle.*

would not break. She began to sing a mournful song. Beneath the rock lay a large black lobster, sleeping quietly after a long battle with an enemy in the sea. The tapping on the rock had awakened him, and he rubbed his eyes and listened to her song of lament. When he understood what she wanted, he remembered his oath at the great gathering of the fish, and he determined to punish her. He came out of his hiding place and, waving his claw politely, asked if he could open the box. But when the girl held the box out to him, the lobster grabbed her by the wrist with his strong claw and swam with her far out to sea. It is believed that the lobster sold the girl to a merman, and that she is still slowly being changed into a fish. It is certain that she never came back to land.

On the first day of May, however, she always appears on the water looking into a mirror to see if she is closer to becoming a fish than she was the year before. She combs her long hair which is now covered with pearls, and she looks with longing eyes toward her old home. Sometimes on moonlit nights, fishermen hear her strange, plaintive song across the water. When they do, they stay on shore, for they know that she is lonely and that she might seize them and carry them off to be her playmates in her home of bright shells far under the sea.

The Mermaid of the Fairy Tale

In the imagination of Hans Christian Andersen the mermaid became a thoroughly lovable girl—a girl who gave up her life in the sea in the hope of winning the love of a mortal and of gaining an immortal soul.

Andersen's little mermaid was the youngest of six girls who lived in a castle of coral with a roof of mussel shells filled with pearls. They were friends of the fish, who ate from their hands and allowed the girls to pet them.

Many an evening the five older sisters rose arm in arm to the surface of the sea to sing, and they had voices more beautiful than any mortal. Whenever a storm was approaching and they thought a ship might be wrecked, they swam ahead of it and sang to the sailors about how beautiful it was at the bottom of the sea. The little mermaid always felt very sad at being left all alone at home, and she looked as though she were going to cry. But she did not cry, because a mermaid has no tears to shed. So she suffered all the more.

When the little mermaid was grown up enough to go to the surface of the sea, she rescued a prince from a shipwreck and fell desperately in love with him. Back at home she pined for him, and one day she made the dangerous journey beyond the roaring maelstroms to the den of the sea witch. Here she made her fateful bargain. In the hope of winning the prince's love, she agreed to give up her life in the sea. But alas, the prince married another. The poor little mermaid was just about to turn into foam when the kindly spirits of the air rescued her.

Hans Christian Andersen's Little Mermaid *did gain the immortality she yearned for, both in the fairy tale and in her statute by the shore in Copenhagen (right). Even today, Scandinavia seems the true home of the mermaid. The painting of her below is by a child of Sweden.*

Chapter VI. The Impossible Mistress

The impossible love of the fisherman for the mermaid is only the legendary illustration of the violent and contradictory feelings man always developed for the sea. Immense, but limited, powerful and vulnerable, prolific but often empty, the sea inspires hope and despair, enthusiasm and rage, peace and terror. She can be beautiful and inviting, but becomes at once terrifying and cruel. She is loved and hated at the same time by most sailors. An impossible mistress for millenia, the sea became our indispensable spouse.

"*Odi et amo* [I hate and I love] may well be the confession of those who consciously or blindly have surrendered their existence to the fascination of the sea."
Joseph Conrad

We knew her better. The study of the oceans and of marine life had become a science, albeit one still in its adolescence. The myths and legends had drifted down to the people and many of the tales had become a part of our cultural heritage. Every child knew who the Greek god of the sea was. He knew the journey of Odysseus and of Jason and the Argonauts. The people knew about Noah and his Ark and Jonah and the Whale. But we needed an art that was more mature, that reflected the advances that had been made in man's knowledge of himself and the world around him.

Man understood his own emotions much better than he had. He knew that he was a very complex being, and the heroes of his new art could be no less complex. The world around him, even the sea, had become more complex, too, as it had become better understood. There was no longer a great river that ran around the earth, but there were many currents in the oceans that were very much like rivers. There were no monsters in the sea or terrible whirlpools to devour men and ships, but that did not mean the sea was without its dangers. There were no nymphs of the sea or water sprites or lovely mermaids, but there was great beauty and charm to be found everywhere in the sea. There was mystery enough in the spawning of the eel. There was sufficient heroism in the lifetime of a salmon. There were countless wonders that were *real.*

The sea, the real sea, began to become a part of our cultural heritage. Benjamin Britten turned the story of Noah and the Flood into a brilliant musical morality play. Ernest Hemingway and Yukio Mishima gave us two of the best novels of the twentieth century in *The Old Man and the Sea* and *The Sound of Waves.* Eugene O'Neill greatly enriched our theater with his plays set at sea. Some of our greatest poets continued to look to the sea for inspiration—men and women like Pablo Neruda, Robinson Jeffers, and Gabriela Mistral. Choreographers, especially Doris Humphrey and Gerald Arpino, incorporated its rhythms and movements in our modern dance and ballet. Painters continued to paint the sea, and Winslow Homer and John Marin rarely departed from it. The photographer and the filmmaker—and even the architect—began to pay increasing attention to it as a source of inspiration. It is intriguing to speculate on what art will be when the sea and its life have become as much a part of our culture as our very soul.

***Havis Amanda,** Finland's most famous sea nymph, stands in the marketplace in Helsinki.*

A ship lies wrecked (above) and a diver searches among the wreckage of another that sank (below). ***Shipwrecks*** *have always been a source of inspiration to the poet. Shakespeare's* The Tempest *opens with a storm and a shipwreck that are among the most marvelous moments in art inspired by the sea.*

Full Fathom Five

Shakespeare thought of the sea much the same as everyone else in England did at the time—as the country's guardian, the barrier between England and her enemies. Usually Shakespeare had only opprobrious terms for it, calling it rough, rude, wayward, dangerous, hungry, ruffian, ruthless. In *The Taming of the Shrew* the sea became a wild boar and the foam its sweat. In *Othello,* Lodovico describes the villain Iago in the worst terms he can think of when he says he is "More fell than anguish, hunger, or the sea!" Romeo associates the sea with hungry tigers as he comes to the tomb where beloved Juliet lies. His intentions are "More fierce and more inexorable far, Than empty tigers, or the roaring sea."

Yet here and there is a suggestion of a certain affection for it, as in *A Midsummer Night's Dream,* where the "dulcet and harmonious breath" of a mermaid makes the rude sea grow civil, or in *The Tempest,* which

begins with a great storm and shipwreck, and which is full of imagery of the sea, in which Ariel has her delightful song:

> Full fathom five thy father lies;
>   Of his bones are coral made;
> Those are pearls that were his eyes;
>   Nothing of him that doth fade
> But doth suffer a sea-change
> Into something rich and strange.
> Sea nymphs hourly ring his knell:
>
> Ding-dong.

*A section from the **Bayeux Tapestry** (below), made between 1066 and 1077 for Odo, bishop of Bayeux and half-brother of William the Conqueror. Its 76 panels tell the story of the Norman Conquest of England. Here William's sailors are about to cross the English Channel.*

Apostrophe to the Ocean

If the poet were not so readily attracted to the sea, surely the sea would make poets of us. From ancient times to the present day the sea has enticed him with its own rhythm and movement, its power and expanse. The poet Byron saw the sea as the source of spiritual renewal for the child of nature who had become disgusted and disillusioned with the world, whose wanderings and reflections he described in his long poem, *Childe Harold's Pilgrimage.*

Roll on, thou deep and dark blue Ocean—
  roll!
Ten thousand fleets sweep over thee in vain;
Man marks the earth with ruin—his control
Stops with the shore—upon the watery
  plain
The wrecks are all thy deed, nor doth remain
A shadow of man's ravage, save his own,
When, for a moment, like a drop of rain,
He sinks into thy depths with bubbling
  groan,
    Without a grave—unknelled,
        uncoffined, and unknown.

The Effect of Sunlight on Water *by the French painter André Derain (1880–1954).*

Thy shores are empires, changed in all save
thee—
Assyria—Greece—Rome—Carthage—
what are they?
Thy waters washed them power while they
were free,
And many a tyrant since; their shores obey
The stranger, slave, or savage; their decay
Has dried up realms to deserts—not so thou;
Unchangeable save to thy wild waves' play;
Time writes no wrinkle on thine azure brow—
Such as Creation's dawn beheld, thou
rollest now.

And I have loved thee, Ocean! and my joy
Of youthful sports was on thy breast to be
Borne like thy bubbles, onward: from a boy
I wantoned with thy breakers—they to me
Were a delight; and if the freshening sea
Made them a terror—'twas a pleasing fear,
For I was as it were a Child of thee,
And trusted to thy billows far and near,
And laid my hand upon thy name—as I
do here.

George Gordon,
Lord Byron

The effect of a cliff on waves *is seen as an onrushing wave meets another on the rebound.*

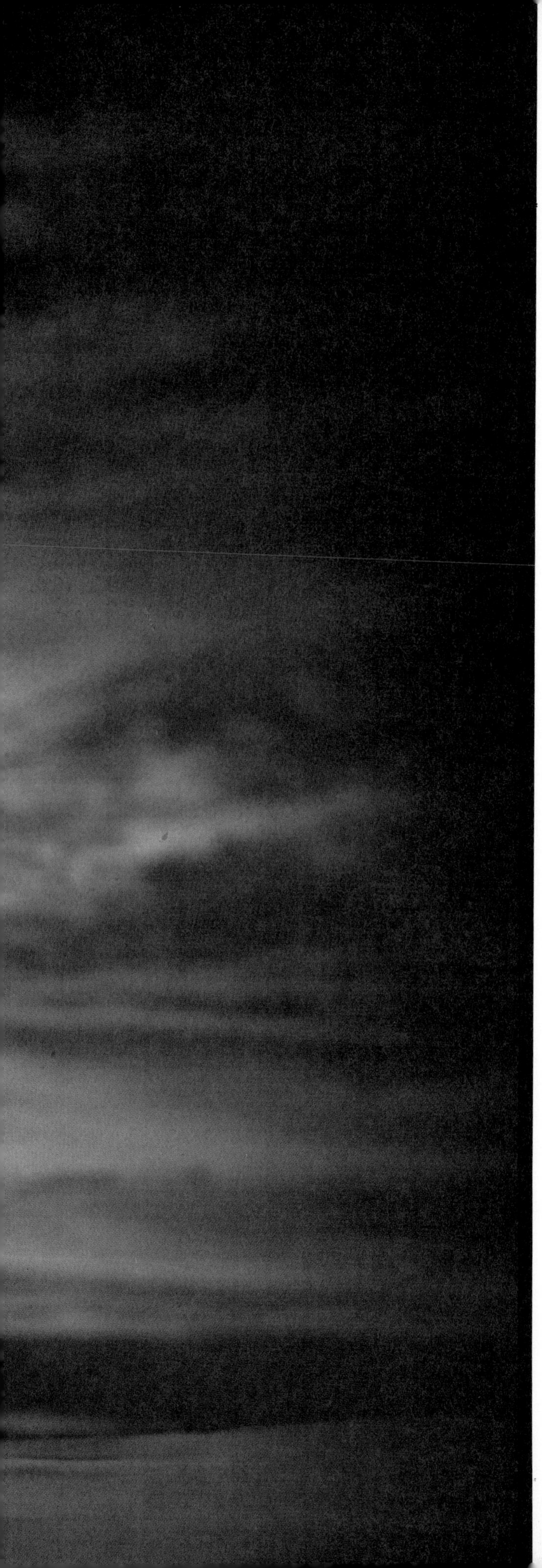

The Flying Dutchman

The opera has, on occasion, taken the sea as its dominant image. Bizet's *The Pearl Fishers* is a familiar example. Benjamin Britten's *Peter Grimes* is also about a fisherman and about life on the sea. The same composer turned to the sea for his *Billy Budd,* derived from the Melville work of the same title. Karl Maria von Weber's *Oberon* has a dramatic aria called "Ocean, Thou Mighty Monster," as the character contemplates the sea's awesome grandeur. Perhaps even better known is Richard Wagner's *The Flying Dutchman.* This work has kept alive a very old legend of the sea.

Wagner composed the opera in 1843 after making a sea voyage that almost ended in disaster. His discomfort on the stormy North Sea filled him with sympathy for the unfortunate Dutch captain in the legend.

The story tells of the Dutchman who is doomed to sail forever on his ship the *Flying Dutchman* until the love of a faithful woman redeems him for his challenge of heaven and hell. Once every seven years he is permitted to go ashore to find that love. During one of these periods, he is driven by a storm to a Norwegian harbor. Here he meets Senta, the woman of his dreams. Senta rejects her former lover to give herself to the Dutchman. Reasoning that if she could be unfaithful to Erik she could also be unfaithful to him, he suspects that his hopes for redemption are again to be shattered. He returns to his ship and, though a storm is raging, sets sail. Senta climbs to the top of a cliff, shouting to the Dutchman that she has always been faithful to him and will be until death. Then she throws herself into the sea. The *Flying Dutchman* immediately vanishes beneath the waves. The Dutchman has after all been redeemed. Embracing, the forms of Senta and the Dutchman rise heavenward.

In Herman Melville's **Moby Dick** *a white whale drove Captain Ahab and the crew of the* Pequod *through all the oceans of the world in its pursuit.*

The Greatest Whale of All

The sea has always been significant in American life, first with its colonies huddled close to the Atlantic shore and later when the nation pushed south to the Gulf of Mexico and then to the Pacific. It was inevitable that literature in the United States should reflect that proximity.

James Fenimore Cooper invented the sea novel in the United States and Herman Melville brought it to perfection. Cooper is probably best known as a writer of the forest frontier, as in *The Last of the Mohicans,* but he wrote a dozen novels of the sea as well. In *Red Rover, Afloat and Ashore, The Sea Lions, Ned Myers, The Water-Witch,* and others, one can see the beginnings of an attempt to make the novel of the sea a vehicle for meanings as universal as birth and death. The man at sea begins to become representative of all men.

In the years shortly after Melville returned from his wanderings in the South Seas, he published two novels based on his experiences there—*Typee* and its sequel *Omoo.* In *Redburn,* which followed soon after, Melville described a voyage across the North Atlantic on a merchant vessel. In *White Jacket* the setting is the cruise of an American man-of-war from the port of Callao in Peru around the Horn and back to its home port on the eastern seaboard. In 1851 he published what is perhaps the greatest novel of the sea ever written—*Moby Dick.*

Moby Dick is a great white whale, and the novel tells how the whale is pursued by the crew of the *Pequod* under Captain Ahab until it destroys the ship and only one man survives. One of Melville's most eloquent passages is this description of the nursing infant whales and their mothers seen by the men in the *Pequod's* whaling boat:

"But far beneath this wondrous world upon the surface, another and still stranger world met our eyes as we gazed over the side. For, suspended in those watery vaults, floated the forms of the nursing mothers of the whales, and those that by their enormous girth seemed shortly to become mothers. The lake, as I have hinted, was to a considerable depth exceedingly transparent; and as human infants while suckling will calmly and fixedly gaze away from the breast, as if leading two different lives at the time, and while yet drawing mortal nourishment, be still spiritually feeding on some unearthy reminiscence—even so did the young of these whales seem looking up towards us, as if we were but a bit of Gulf-weed in their new-born sight. Floating on their sides, the mothers also seemed quietly eyeing us. One of these little infants that from certain queer tokens seemed hardly a day old, might have measured some fourteen feet in length and some six feet in girth. He was a little frisky; though as yet his body seemed scarce yet recovered from that irksome position it had so lately occupied in the maternal reticule, where, tail to head, and all ready for the final spring, the unborn whale lies bent like a Tartar's bow. The delicate side-fins, and the palms of his flukes still freshly retained the plaited crumbled appearance of a baby's ears newly arrived from foreign parts."

Of Moby Dick himself, Melville writes that "a gentle joyousness—a mighty mildness of repose in swiftness, invested the gliding whale. Not the white bull Jupiter swimming away with ravished Europa clinging to his graceful horns; his lovely, leering eyes sideways intent upon the maid; with smooth bewitching fleetness, rippling straight for the nuptial bower in Crete; not Jove, not that great majesty supreme! did surpass the glorified White Whale as he so divinely swam."

Moby Dick was never caught. *At the end of the novel the whale capsizes the* Pequod *and all except Ishmael are drowned.*

It is a novel that can be taken simply as the story of a single American whaling voyage, in which there are descriptions of the whale, whale burning, and the whole gory process aboard a ship that reduces once-proud whales to barrels of oil in the hold. But it is also a novel that can be regarded as an evocation of the whole human drama. Perhaps most importantly, the novel is one that can help us to understand that life can only be honestly confronted in the loneliness of each human heart and that it is in a relation to the sea that we might best know that loneliness. It also suggests that there is no better place to search for a meaning to existence than in the inscrutable sea, where everything is in eternal flux.

D. H. Lawrence wrote of Melville: "The greatest seer and poet of the sea for me is Melville. His vision is more real than Swinburne's, because he doesn't personify the sea, and far sounder than Joseph Conrad's, because Melville doesn't sentimentalize the ocean and the sea's unfortunates."

The Sea of Joseph Conrad

Few of the many poets, playwrights, short-story writers, and novelists who have written about the sea have known it realistically except, perhaps, from the vantage of the shore. From Homer and Shakespeare to Coleridge, Heine, Baudelaire, Mallarmé, Rimbaud, Whitman, Hopkins, and Perse, they have made their poetic resources and insights a substitute for actual experience. A few, like Marryat and Cooper, Dana and Melville, were exceptions. So, too, was Joseph Conrad.

The sea and seafaring life that Conrad gave us in *The Mirror of the Sea, The Nigger of the "Narcissus," The Shadow-Line, Youth, Typhoon,* and elsewhere was born of an intimate knowledge of that reality. At seventeen he went to sea, and he remained a sailor for twenty years. The experience made him realize that the sea was not easy to love.

"For all its fascination that has lured so many to a violent death, its immensity has never been loved as the mountains, the plains, the desert itself, have been loved. Indeed, I suspect that, leaving aside the protestations and tributes of writers who, one is safe in saying, care for little else in the world than the rhythm of their lines and the cadence of their phrase, the love of the sea, to which some men and nations confess so readily, is a complex sentiment wherein pride enters for much, necessity for not a little, and the love of ships—the untiring servants of our hopes and our self-esteem—for the best and most genuine part."

For Joseph Conrad, the sea and woman were in many ways almost one. In *The Arrow of Gold,* Monsieur George, who is strongly reminiscent of the young Conrad, says: "Woman and the sea revealed themselves to me altogether, as it were: two mistresses of life's values. The illimitable greatness of the one, the unfathomable seduction of the other working their immemorial spells from generation to generation fell upon my heart at last, a common fortune, an unforgettable memory of the sea's formless might and of the sovereign charm in that woman's form wherein there seemed to beat the pulse of divinity rather than blood."

In ***Hawaiian mythology,*** *it is said that Hina-opuhala-koa is the goddess of coral and all the spiny creatures of the sea. Sometimes she appears as a woman and sometimes as a coral reef.*

The Chantey

Workers who labor at a common task often need a song to lighten their burden. It is a shame that the chantey, as such a song is called, is not heard now so often as it once was. There was a day when every lumberjack and railroad worker, every worker on the dock and every roustabout at the circus had his own song to sing. And, of course, the sailors had theirs. While hoisting sail or hauling anchor, the leader, the chanteyman, sang the verse and all the sailors joined in the chorus. The rhythm helped them coordinate their efforts. The songs they sang might be about any of the many thoughts that occupy a sailor's mind, and they are often perceptive and humorous.

Oh, a ship was rigg'd, and ready for sea,
And all of her sailors were fishes to be.

Windy weather! Stormy weather!
When the wind blows we're all together.
Blow, ye winds, westerly, gentle
southwesterly,
Blow, ye winds, westerly—steady she goes.

Oh, first came the herring, the king of the sea,
He jumped on the poop, "I'll be captain,"
said he.

The next was a flatfish, they call him the
skate,
"If you be the captain, why, sure, I'm the
mate."

The next came the hake, as black as a rook,
Says he, "I'm no sailor, I'll ship as the cook."

The next came the shark, with his two rows
of teeth:
"Cook, mind the cabbage, and I'll mind the
beef."

And then came the codfish, with his
chucklehead,
He jumped in the chains, began heaving the
lead.

The next came the flounder, as flat as
the ground:
"Chucklehead, damn your eyes, mind how
you sound."

The next comes the mack'rel, with his
striped back,
He jumped to the waist for to board the
main tack.

And then came the sprat, the smallest of all,
He jumped on the poop, and cried,
"Main topsail haul."

Windy weather! Stormy weather!
When the wind blows we're all together.
Blow, ye winds, westerly, gentle
southwesterly,
Blow, ye winds, westerly—steady she goes.

When God sent the flood and Noah took his family and all the animals in pairs aboard the ark, the fish mocked God for his powerlessness over them. In order to show the fish that this was not so, God brought his fist down upon them and flattened them with one blow. Some fish, like the ***flatfish*** *(left), got flattened more than others.*

Folklore of the Pacific Marshall Islands. *The unfaithful wife of a chief took an eel (below) as her lover. When the husband found them together, she poured coconut oil on her paramour, and it slithered through the hands of its would-be captor. The eel escaped into the first hole it came to because now it was so slippery it could go anyplace. And so the legend concludes that is why all eels are very slippery and never stay in the same hole.*

Music and the Sea

The imagery that the sea can provide has always had an appeal to the composer. The sound of the sea itself as it is most familiar to us—in the steady roll of waves onto a shore—provides us with one of the most fundamental rhythms that we know.

The nineteenth century Russian pianist and composer Anton Rubenstein called his second symphony *The Ocean*. The English composer Sir Edward Elgar wrote a cycle of songs for contralto with orchestral accompaniment called "Sea Slumber Song," "In Haven," "Sabbath Morning at Sea," "Where Corals Lie," and "The Swimmer." The songs were first sung in 1899. Ralph Vaughan Williams, another great English composer, turned to the sea for inspiration on several occasions. His *Sea Symphony,* first performed in 1910 is the most familiar of these works. *Sea Symphony* is a choral work that makes use of Walt Whitman's poem "Sea Drift" in its first three movements. In the finale, the sea is at first forgotten, and the character of the poet sets out on a quest for the meaning of life. But his soul, in surveying the universe, keeps returning to the sea as the symbol of its journeys:

"O we can wait no longer
We too take ship, O Soul.
Joyous we too launch out on trackless seas"

and at the end

"Away O Soul, hoist instantly the anchor"

till the voices die away in

"O farther, farther sail."

Perhaps the most familiar piece of music suggested by the sea is *La Mer* by Claude Debussy. Debussy began to write this great work in 1902 and completed it three years later. His letters of the time show that his childhood impressions of the Mediterranean

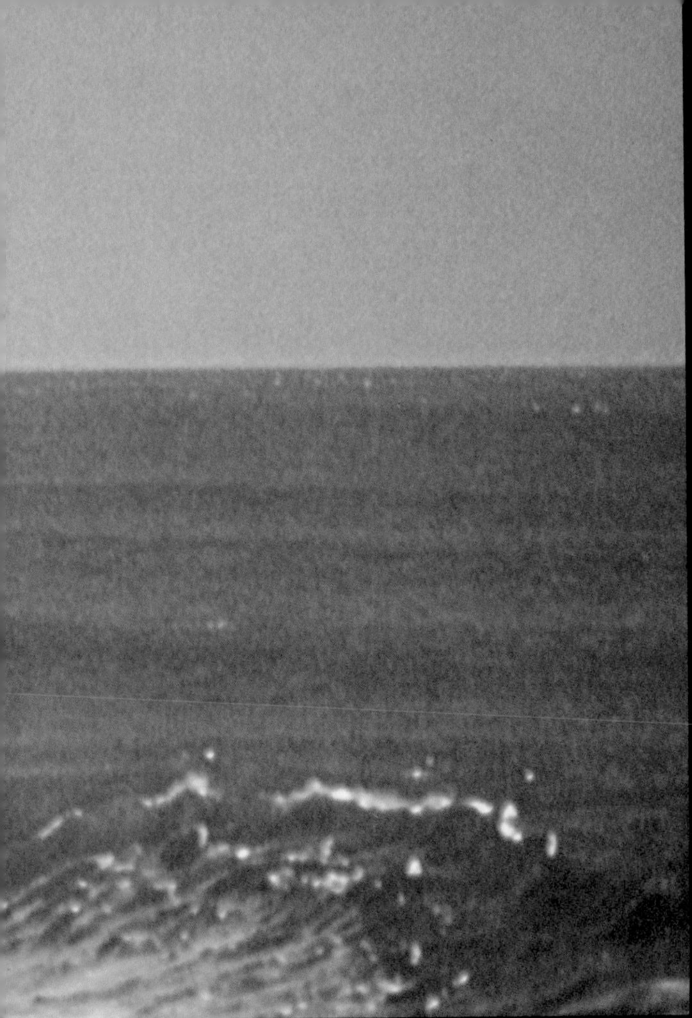

at Cannes and of the career of a sailor, once planned for him by his father, were still very much alive. Subsequent summers spent on the coast of Normandy and in the Channel Islands enriched the young musician's experience of the sea. "You perhaps do not know that I was destined for the fine life of a sailor and that it was only by chance that I was led away from it. But I still have a great passion for the sea," he wrote in a letter in 1903. In *La Mer*—with its suggestions of gleaming spray, the crash of waves, and the gurgling backwash—there are passages as moving and expressive as any ever written by an impressionistic composer. The passage at the end of the first movement is surely the greatest evocation of nature in a work for orchestra. The rocking barcarole rhythm of the strings leads to foreboding chords on the horns accompanied by eerie patterns played by the harp. A great pantheistic drama of the sea presents itself.

The Sea in Dance

It should not be surprising that the sea, and especially the rhythmical movement of the sea, has been an inspiration to the choreographer: Mary Anthony's *Threnody*, Frederick Ashton's *Ondine*, and John Cranko's *Pineapple Poll* are a few of the story ballets that derive their principal imagery from the sea.

In Birgit Cullberg's *Lady from the Sea*, suggested by Henrik Ibsen's play of the same name, Ellida loves a sailor who leaves her and returns to the sea. She marries a widower, the father of two daughters, but she feels miserable in her new home. When the sailor returns, her husband offers Ellida her freedom, but she does not leave him. She realizes that the sailor, like the sea, represents for her only a romantic yearning.

In 1963 Gerald Arpino gave us his *Sea Shadow*, a dramatic pas de deux in which a sprite from the sea visits a dreaming boy on a beach. Attracted as the boy is to the nymph, he soon understands that she will lead him not only to happiness but also to his death. It is one of the most beautiful duets in all ballet.

Perhaps the perfect example of a presentation of the sea in dance is Doris Humphrey's *Water Study*. In this ballet without music, the dancers—12 girls—imitate the undulation of the waves and the movement of the surf onto and from the shore. To feel the sea with one's whole body, as these dancers must, and as the audience is asked to do, is to understand something very fundamental about it in a way that can mean more than all the books one might read.

A boy dreaming on a beach is visited by a nymph of the sea in the ballet **Sea Shadow.** *As our awareness of the sea reawakens, the ballet, like all the arts, will make more and more use of its imagery.*

Giradoux's Sea Nymph

Jean Giradoux made the mermaid a creature of our time when he wrote *Ondine*. Ondine is a 16-year-old sea nymph who had been left by the side of a lake as an infant in exchange for the child of a fisherman and his wife. Ondine is gifted in magic, but she owes perfect obedience to the Old Man of the Sea, with whom she makes a pact. She falls in love with a mortal, a knight, and she agrees to be allowed to love him on the condition that if he deceives her he will die.

The other nymphs are jealous of Ondine and say that she let herself be taken too easily. Ondine reminds one of them that "a narwhal drizzled you with his jet of water, and you gave yourself to him without a word."

The knight, Hans, tells his beloved mermaid that people who live on land and love each other do not stay together all the time. Ondine is appalled at this and tells the knight of the dogfish couple—so long as they live they stay together. "Through storm and calm they swim together, thousands and thousands of miles, side by side, two fingers apart, as if an invisible link held them together. . . . As long as they live, not even a sardine can come between them."

Years later both Ondine and the knight deceive each other. She disappears, but the knight finds a wreath of starfish and sea urchins on the chapel door and so knows she can not have gone far. Then one day a fisherman catches her in his net. The two lovers are brought together for a last farewell. The knight dies and Ondine loses all memory of her sojourn on land and returns to her home in the depths of the sea.

Because of reflecting sunlight playing on the water's surface, the ***eye of a fish*** *reminds us of that thin barrier that separates creatures of the air from those of the sea.*

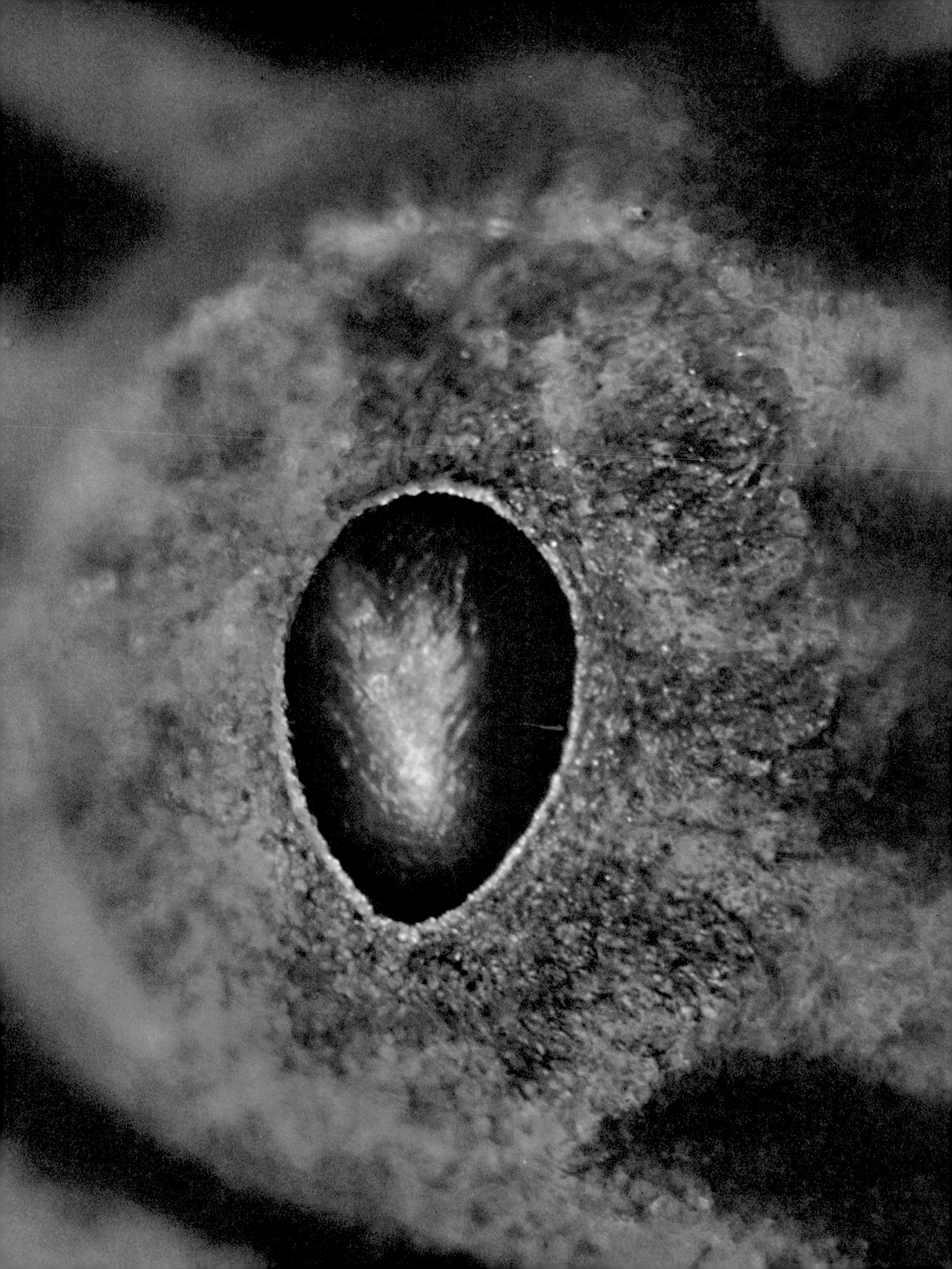

Man Fishing

"It is enough that we do not have to try to kill the sun or the moon or the stars. It is enough to live on the sea and kill our true brothers." These are the words of Santiago, the fisherman, in Ernest Hemingway's short novel *The Old Man and the Sea*.

The old man has gone for 84 days without catching a fish. He is unlucky, and he has only the youth, Manolin, a boy lovingly described as "already a man" in his knowledge of things of the sea, to give him encouragement. But Santiago sets out alone in his skiff and sails out far beyond the other fishermen. He is alone on the Gulf Stream, until he sees a flight of wild ducks go over. Then he remembers that "no man was ever alone on the sea." Of the two porpoises that frolic near the skiff he says, "They are good. They play and make jokes and love one another. They are our brothers like the flying fish."

The marlin that he catches is both his "brother" and his "friend," and all the qualities he sees in the fish—beauty, nobility, courage, calmness, and endurance—are the qualities he values most.

But both Santiago and the marlin have enemies in the sea. Tied along the side of the boat, the marlin is attacked first by a mako shark, one of the biggest and most dangerous sharks. Then, as the skiff sails toward home, come the shovel-nosed sharks in packs, and they leave nothing but the skeleton of the great fish for the old man to bring to shore.

But he has not been defeated. The old man's triumph is not so much in his endurance in having put up a good fight or even in his acceptance of the loss of the fish but in his understanding of man's kinship with all the other creatures of the world. Santiago is not simply a fisherman. He is Man Fishing.

He always thought of the sea as la mar *which is what people call her in Spanish when they love her. Sometimes those who love her say bad things of her but they are always said as though she were a woman.*

Hemingway, The Old Man and the Sea

Chapter VII. Reconciliation

Once the sea belonged to no one, and no one took care of it. The sea was the bailiwick of the gods, and their responsibility.

That was in the childhood of man. When he grew older and wiser, he came to know that the sea was as necessary to him as the air he breathed or the food on his table, and he knew that his gods had failed him. The sea was in danger, and if the sea—and man—were to survive, man alone must take on the responsibility he had shirked for so long.

"I am indeed lord of the world, but the law is lord of the sea."
Antonius,
Emperor of Rome

The seas had belonged to no one, and now they belonged to everyone, even to those 29 nations that do not touch the sea at all. But few men understood this.

There had been more regulations in effect in antiquity than there were now. In Roman law it was stated that "By the law of nature . . . the following things are common to all men: the air, flowing water, the sea, and, consequently, the shores of the sea." Centuries later Louis XIV of France decreed that all the beaches of France should forever be accessible to all. That royal decree has been upheld to this day. Cicero, the Roman poet and statesman, described the principle of common use of the sea with these glowing words: "This, then, is the most comprehensive bond which unites together men as men and all to all; and under it is the common right to all things that nature has produced for the common use of man . . ."

From the time of Rome's decline, a theory of maritime freedom continued to be built based on the Roman model. Vasquez, the great Spanish jurist of the early sixteenth century, maintained that to make the seas private national property was contrary to the law of nature and the basic principles of international relations. Then came the eloquent formulation of Hugo Grotius.

The law of the sea is at present in a state of upheaval. In 1945 a dangerous precedent that for the first time posed a serious threat to the concept of freedom of the seas that had taken many centuries to formulate. The Truman Proclamation declared that national sovereignty extended to the resources of the continental shelf. Other nations quickly followed suit, claiming jurisdiction over an area as much as 200 miles from shore. Now we are faced with the possibility of unmitigated chaos on and below the seas as nations scramble to expand their claims and, in one way or another, to establish dominion over the ocean. The nations with the technological capacity do not need to make a formal claim. They exercise an unspoken claim in their capacity to exploit its wealth at the expense of others.

In 1958 the Geneva Convention brought us one step closer to chaos when it declared the high seas a no-man's-land, belonging to no one, with no one in control.

It would be a great shame for all men if now, through selfishness or simply by not caring, we failed to achieve a reconciliation with the ocean and among ourselves.

Conflict. *Nations that have been traditional allies and partners in social and economic cooperation still fall into violent disputes over territorial claims at sea. The oceans could be the starting place of another world war or they could be the place where a reconciliation among nations begins.*

17

The Fishing Factories

One of the oldest and most economically significant uses of the sea is fishing. Today there exist serious threats to the depletion of stocks through overfishing. Two-hundred-thousand-ton fishing and whaling factories belonging to a few of the advanced nations make enormous catches, while developing nations must continue to rely on age-old methods to feed their people from the sea.

Traditional international law can no longer keep any kind of equitable distribution in the face of technological advances in ship construction and fishing methods. The growing tendency of nations to expand their exclusive national fishery zones—in some cases as far as 200 miles from the coast—is one of the many grave encroachments on the traditional freedom of the seas. At present there is an almost total absence of effective international regulation, let alone cooperation, in the management and distribution of fisheries. Successive international whaling agreements concluded between 1931 and 1946 have not prevented ruthless overkill by the world's major whaling fleets and the near extinction of the blue and other whales.

The living resources of the sea should no longer be regarded as available to the first arrival to catch or destroy at will, but a treasure to be managed as fairly as possible under some kind of international supervision. Instead of being no man's land, the sea belongs to all mankind.

The exploration of the ***seabed and ocean floor*** *and the exploitation of its resources shall be carried out for the benefit of mankind as a whole, irrespective of the geographic location of States, whether landlocked or coastal, and taking into particular consideration the interests and needs of the developing countries . . .*

From the United Nations
Resolution of December 17, 1970

Energy and Defense

Advanced nations are facing a disastrous energy crisis which has sent them in search of new sources of oil and gas in the seabed, which is unbelievably rich in both resources. More and more efficient methods of exploration and exploitation of these deposits and of mineral deposits have been developed. Seabed oil production, which amounted to a scant one percent of production by the United States in 1956 had become 17 percent by 1970, and seabed oil, already 20 percent of the total world production, may reach as high as 33 percent by 1980.

The most disputed portion of the seabed is the outer continental margins, and here it has been estimated that there is more than a trillion dollars worth of oil, gas, and mineral resources of many kinds. But at present we have still failed to work out an international agreement that will guarantee the fair use of these riches.

Compounding the present problems of the sea is its use for defense. Since 1945 ocean

armament has become the heart of superpower weaponry. Submarines had already become a significant factor in naval warfare in World War I, and by the end of World War II they were a principal factor. It was clear that in the future it would be the submarine that would control the seas. Now several nations possess nuclear-powered submarines capable of great speeds and submersion for considerable lengths of time, and many of these are equipped to deliver multiple nuclear warheads from missiles. It is undoubtedly the most perverse use of the sea that has ever been devised.

Crew members *of the nuclear submarine* Skate *take some fresh arctic air.* Skate *was the first submarine to surface through ice at the North Pole. The* Nautilus *had traveled submerged from the Pacific to the Atlantic in 1958.*

At present the predominant technique of ***exploiting offshore oil or gas*** *is from a fixed platform supported by piles fixed in the sea floor (below). As operations proceed to greater depths, this method will become increasingly uneconomical.*

Common Heritage of Mankind

On December 17, 1970, the General Assembly of the United Nations adopted a "Declaration of Principles Governing the Seabed and the Ocean Floor, and the Subsoil Thereof, Beyond the Limits of National Jurisdiction." It declared that the seabed and the ocean floor, and the subsoil of both beyond the limits of national jurisdiction as well as the resources of the area, are the "common heritage of mankind." It declared that the area should not be subject to appropriation by any means either by states or by persons, natural or juridicial, and that no state should claim or exercise sovereignty or sovereign rights over any part of it. It declared that an international regime was to be established with appropriate machinery to give effect to these provisions. The regime would provide for the orderly and safe development and rational management of the sea and its resources.

At the same time the General Assembly voted to convene a comprehensive Conference on the Law of the Sea in 1973 to deal with a precise delimitation of the seabed area, a regime for the high seas, territorial waters, international straits, the continental shelf, fishing, contiguous zones, the preferential rights of coastal states, and the prevention of marine pollution.

It was only a glimmer of hope, still awaiting more practical measures, but it was our best one, because we knew that no unilateral agreements would ever suffice and that only through an agreement that all nations signed could the future of the sea be secure.

*The **seabed and ocean floor,** and the subsoil thereof, beyond the limits of national jurisdiction ... as well as the resources of the area, are the common heritage of mankind ...*

From the United Nations
Resolutions of December 17, 1970

“And God said unto Noah, ‘This is the token of the covenant, which I have established between me and all flesh that is upon the earth.’ ”

GENESIS

Spirit of the Sea

Beyond the mountain droves
Lowing in morning dust
Away from social caves
Elsewhere by fantasy
In arcana of Mind
Rave the dreams of the Sea
To caress the soft waves
All Souls spread out their wings

Canticles from sunken temples
Resound and rise in autumn mists

Tollings of spectral bells
The laughing of undines
Hulks of epic galleys
And Gods of verdigris
Slowly engulfed in mud
Conceal man's infancy

Skeletal myths
Vapors of fear
Freedom dances
Fevers of risk

Coin foster and fondle
The Poet of the Sea.

Index

ILLUSTRATIONS AND CHARTS:

Howard Koslow—22-23, 43.

PHOTO CREDITS:

Maria Arvidsson—108; Gail Ash—5; The Bettman Archive Inc.—78, 79, 118, 119; Bruce Coleman Inc.: Jen and Des Bartlett—131, J. Dermid—40-41, Allan Power—99, Sieger E. Wilhelm—122; James Dugan—74; European Art Color Slide, Peter Adelberg, Inc.—12-13, 14, 16, 36, 76-77, 87, 113 (bottom); Freelance Photographers Guild: Dennis L. Crow—17, FPG—46-47, L. Grigg—98, Dennis Hallinan—121, Phil Lustig—28, Orion Press—84-85, Alan B. Seiden—67, Charles Serra—116-117, Robert Simmons—137 (right), Western Marine Laboratories—123; Henry Genthe—15; Information Office of the City of Helsinki—111; Charles Mather—130; Herbert Migdoll—126, 127; Richard C. Murphy—124-125 (top), 142; Musée de l'Annonciade (Permission ADAGP 1973 by French Reproduction Rights Inc., Photo Henri Tabah, Permission SPADEM 1973 by French Reproduction Rights)—114; Museum fur Volkerkunde und Schweizersches Museum fur Volkskunde Basel (photo by Colorophoto Hanz Hinz)—96; Naval Photographic Center—60-61, 136-137 (left); William Rockhill Nelson Gallery of Art, Kansas City, Mo. (photo by Myron Wang)—20, 53; Nippon Suisan Company, Tokyo—104, 105; Paris Bibliothèque Nationale, 75, (photo by Lauros-Giraudon); Paris Musée de la Marine (photo by Lauros-Giraudon)—54-55; Philadelphia Museum of Art: The Louise and Walter Arensberg Collection (photo by Alfred Wyatt)—90-91; Carl Roessler—92-93 (top); Royal Viking Line—66; David Schwimmer—109; The Sea Library: Francois R. Brenot—97, The British Museum—62, Jack Drafahl—94, 100-101, 128-129, Marx—68, 69, Emerson Mulford—81, Chuck Nicklin—120, Carl Roessler—29, The Sea Library—11, 18, 19, 33, 48, 63, 103, Phillip Drennon Thomas—24, 25, 38-39 (top), 49, 50, 80, 106-107, Joseph A. Thompson—57, UCLA Art Department—35; Tom Stack & Associates: American Stock Photos—70-71, Larry C. Moon—26-27, 34, 115, 125, (bottom), Darrel Ward—140-141; Surfer Publications Inc.: Jeff Devine—112-113 (top); Taurus Photos: P. E. Baker—82, 83, Peter A. Lake—93 (bottom), 134-135, Taurus—102; U.S. Committee for UNICEF: Mohammed Hussan—58-59, Gema Sanchez Lopez—44-45, Lilibet Sikiaridi—88-89; The Winnipeg Art Gallery (G-60.70)—31 (bottom).